CHEMOMETRIC

TECHNIQUES

for QUANTITATIVE

ANALYSIS

CHEMOMETRIC

TECHNIQUES

for QUANTITATIVE

ANALYSIS

RICHARD KRAMER

Aprotec, Inc.
Sharon, Massachusetts

MARCEL DEKKER, INC.   NEW YORK · BASEL · HONG KONG

Library of Congress Cataloging-in-Publication Data

Kramer, Richard.
    Chemometric techniques for quantitative analysis / Richard Kramer.
        p. cm.
    Includes bibliographical references and index.
    ISBN: 0-8247-0198-4
    1. Chemistry, Analytic—Quantitative—Statistical methods. I. Title.
    QD101.2.K73 1998
    543'.0072—dc21   98-4224
                                                                            CIP

This book is printed on acid-free paper.

Headquarters
Marcel Dekker, Inc.
270 Madison Avenue, New York, NY 10016
tel: 212-696-9000; fax: 212-685-4540

Eastern Hemisphere Distribution
Marcel Dekker AG
Hutgasse 4, Postfach 812, CH-4001 Basel, Switzerland
tel: 44-61-261-8482; fax: 44-61-261-8896

World Wide Web
http://www.dekker.com

The publisher offers discounts on this book when ordered in bulk quantities. For more information, write to Special Sales/Professional Marketing at the headquarters address above.

Current printing (last digit):
10 9 8 7 6 5 4 3 2 1

PRINTED IN THE UNITED STATES OF AMERICA

Preface

The proliferation of sophisticated instruments which are capable of rapidly producing vast amounts of data, coupled with the virtually universal availability of powerful but inexpensive computers, has caused the field of chemometrics to evolve from an esoteric specialty at the perhiphery of Analytical Chemistry to a required core competency.

This book is intended to bring you quickly "up to speed" with the successful application of Multiple Linear Regressions and Factor-Based techniques to produce quantitative calibrations from instrumental and other data: Classical Least-Squares (CLS), Inverse Least-Squares (ILS), Principle Component Regression (PCR), and Partial Least-Squares in latent variables (PLS). It is based on a short course which has been regularly presented over the past 5 years at a number of conferences and companies. As such, it is organized like a short course rather than as a textbook. It is written in a conversational style, and leads step-by-step through the topics, building an understanding in a logical, intuitive sequence.

The goal of this book is to help you understand the procedures which are necessary to successfully produce and utilize a calibration in a production environment; the amount of time and resources required to do so; and the proper use of the quantitative software provided with an instrument or commercial software package. This book is not intended to be a comprehensive textbook. It aims to clearly explain the basics, and to enable you to critically read and understand the current literature so that you may further explore the topics with the aid of the comprehensive bibliography.

This book is intended for chemists, spectroscopists, chromatographers, biologists, programmers, technicians, mathematicians, statisticians, managers, engineers; in short, anyone responsible for developing analytical calibrations using laboratory or on-line instrumentation, managing the development or use of such calibrations and instrumentation, or designing or choosing software for the instrumentation. This introductory treatment of the quantitative techniques requires no prior exposure to the material. Readers who have explored the topics but are not yet comfortable using them should also find this book beneficial. The data-centric approach to the topics does not require any special mathematical background.

I am indebted to a great many people who have given generously of their time and ideas. Not the least among these are the students of the short course upon which this book is based who have contributed their suggestions for improvements in the course. I would especially like to thank Alvin Bober who provided the initial encouragement to create the short course, and Dr. Howard Mark, whose discerning eye and sharp analytical mind have been invaluable in helping eliminate errors and ambiguity from the text.

Richard Kramer

Contents

about the author . . .

RICHARD KRAMER is President of Applied Chemometrics, Inc. a chemometrics software, training, and consulting company, located in Sharon, Massachusetts. He is the author of the widely used Chemometrics Toolbox software for use with MATLAB™ and has over 20 years' experience working with analytical instrumentation and computer-based data analysis. His experience with mid- and near-infrared spectroscopy spans a vast range of industrial and process monitoring and control applications. Mr. Kramer also consults extensively at the managerial level, helping companies to understand the organizational and operational impacts of deploying modern analytical instrumentation and to institute the procedures and training necessary for successful results.

This book is based upon his short course, which has been presented at scientific meetings including EAS, PITTCON, and ACS National Meetings. He has also presented expanded versions of the course in major cities and on-site at companies and educational organizations.

Mr. Kramer may be contacted at Applied Chemometrics, Inc., PO Box 100, Sharon, Massachusetts 02067 or via email at kramer@chemometrics.com.

CHEMOMETRIC

TECHNIQUES

for QUANTITATIVE

ANALYSIS

I would give whole worlds to know. This solemn, this awful mystery has cast a gloom over my whole life.
—*Mark Twain*

Introduction

Chemometrics, in the most general sense, is the art of processing data with various numerical techniques in order to extract useful information. It has evolved rapidly over the past 10 years, largely driven by the widespread availability of powerful, inexpensive computers and an increasing selection of software available off-the-shelf, or from the manufacturers of analytical instruments.

Many in the field of analytical chemistry have found it difficult to apply chemometrics to their work. The mathematics can be intimidating, and many of the techniques use abstract vector spaces which can seem counterintuitive. This has created a "barrier to entry" which has hindered a more rapid and general adoption of chemometric techniques.

Fortunately, it is possible to bypass the entry barrier. By focusing on data rather than mathematics, and by discussing practicalities rather than dwelling on theory, this book will help you gain a rigourous, working familiarity with chemometric techniques. This "data centric" approach has been the basis of a short course which the author has presented for a number of years. This approach has proven successful in helping students with diverse backgrounds quickly learn how to use these methods successfully in their own work.

This book is intended to work like a short course. The material is presented in a progressive sequence, and the tone is informal. You may notice that the discussions are paced more slowly than usual for a book of this kind. There is also a certain amount of repetition. No apologies are offered for this—it is deliberate. Remember, the purpose of this book is to get you past the "entry barrier" and "up-to-speed" on the basics. This book is not intended to teach you "everything you wanted to know about ...." An extensive bibliography, organized by topic, has been provided to help you explore material beyond the scope of this book. Selected topics are also treated in more detail in the Appendices.

Topics to Cover

We will explore the two major families of chemometric quantitative calibration techniques that are most commonly employed: the Multiple Linear Regression (MLR) techniques, and the Factor-Based Techniques. Within each family, we will review the various methods commonly employed, learn how to develop and test calibrations, and how to use the calibrations to estimate, or predict, the properties of unknown samples. We will consider the advantages and limitations of each method as well as some of the tricks and pitfalls associated with their use. While our emphasis will be on quantitative analysis, we will also touch on how these techniques are used for qualitative analysis, classification, and discriminative analysis.

Bias and Prejudices — a Caveat

It is important to understand that this material will not be presented in a theoretical vacuum. Instead, it will be presented in a particular context, consistent with the majority of the author's experience, namely the development of calibrations in an industrial setting. We will focus on working with the types of data, noise, nonlinearities, and other sources of error, as well as the requirements for accuracy, reliability, and robustness typically encountered in industrial analytical laboratories and process analyzers. Since some of the advantages, tradeoffs, and limitations of these methods can be data and/or application dependent, the guidance in this book may sometimes differ from the guidance offered in the general literature.

Our Goal

Simply put, the main reason for learning these techniques it to derive better, more reliable information from our data. We wish to use the information content of the data to understand something of interest about the samples or systems from which we have collected the data. Although we don't often think of it in these terms, we will be practicing a form of pattern recognition. We will be attempting to recognize patterns in the data which can tell us something useful about the sample from which the data is collected.

Data

For our purposes, it is useful to think of our measured data as a mixture of Information plus Noise. In a ideal world, the magnitude of the Information would be much greater than the magnitude of the Noise, and the Information in the data would be related in a simple way to the properties of the samples from which the data is collected. In the real world, however, we are often forced to work with data that has nearly as much Noise as Information or data whose Information is related to the properties of interest in complex way that are not readily discernable by a simple inspection of the data. These chemometric techniques can enable us to do something useful with such data.

We use these chemometric techniques to:

1. Remove as much Noise as possible from the data.
2. Extract as much Information as possible from the data.
3. Use the Information to learn how to make accurate predictions about unknown samples.

In order for this to work, two essential conditions must be met:

1. The data must have information content.
2. The information in the data must have some relationship with the property or properties which we are trying to predict.

While these two conditions might seem trivially obvious, it is alarmingly easy to violate them. And the consequences of a violation are always unpleasant. At best it might involve writing off a significant investment in time and money that was spent to develop a calibration that can never be made to work. At worst, a violation could lead to an unreliable calibration being put into service with resulting losses of hundreds of thousands of dollars in defective product, or, even worse, the endangerment of health and safety. Often, this will "poison the waters" within an organization, damaging the credibility of chemometrics, and increasing the reluctance of managers and production people to embrace the techniques. Unfortunately, because currently available computers and software make it so easy to execute the mechanics of chemometric techniques without thinking critically about the application and the data, it is all too easy to make these mistakes.

Borrowing a concept from the aviation community, we can say with confidence that everyone doing analytical work can be assigned to one of two categories. The first category comprises all those who, at some point in their careers, have spent an inordinate amount of time and money developing a calibration on data that is incapable of delivering the desired results. The second category comprises those who *will*, at some point in their careers, spend an inordinate amount of time and money developing a calibration on data that is incapable of delivering the desired measurement.

This author must admit to being a solid member of the first category, having met the qualifications more than once! Reviewing some of these unpleasant experiences might help you extend your membership in the second category.

Violation 1. —Data that lacks information content

There are, generally, an infinite number of ways to collect meaningless data from a sample. So it should be no surprise how easy it can be to inadvertently base your work on such data. The only protection against this is a hightened sense of suspicion. Take nothing for granted; question everything! Learn as much as you can about the measurement and the system you are measuring. We all learned in grade school what the important questions are — who, what, when, where, why, and how. Apply them to this work!

One of the most insidious ways of assembling meaningless data is to work with an instrument that is not operating well, or has presistent and excessive drift. Be forewarned! Characterize your instrument. Challenge it with the full range of conditions it is expected to handle. Explore environmental factors, sampling systems, operator influences, basic performance, noise levels, drift, aging. The chemometric techniques excel at extracting useful information from very subtle differences in the data. Some instruments and measurement techniques excel at destroying these subtle differences, thereby removing all traces of the needed information. Make sure your instruments and techniques are not doing this to *your* data!

Another easy way of assembling a meaningless set of data is to work with a system for which you do not understand or control all of the important parameters. This would be easy to do, for example, when working with near infrared (NIR) spectra of an aqueous system. The NIR spectrum of water changes with changes in pH or temperature. If your measurements were made without regard to pH or temperature, the differences in the water spectrum could easily destroy any other information that might otherwise be present in the spectra.

Violation 2. —Information in the data is unrelated to the property or properties being predicted

This author has learned the hard way how embarassingly easy it is to commit this error. Here's one of the worst experiences.

A client was seeking a way to rapidly accept or reject certain incoming raw materials. It looked like a routine application. The client has a large archive of acceptable and rejectable examples of the materials. The materials were easily measured with an inexpensive, commercially available instrument that provided excellent signal-to-noise and long-term stability. Calibrations developed with the archived samples were extremely accurate at distinguishing good material from bad material. So the calibration was developed, the instrument was put in place on the receiving dock, the operators were trained, and everyone was happy.

After some months of successful operation, the system began rejecting large amounts of incoming materials. Upon investigation, it was determined that the rejected materials were perfectly suitable for their intended use. It was also noticed that all of the rejected materials were provided by one particular supplier. Needless to say, that supplier wasn't too happy about the situation; nor were the plant people particularly pleased at the excessive process down time due to lack of accepted feedstock.

Further investigation revealed a curious fact. Nearly all of the reject material in the original archive of samples that were used to develop the calibration had come from a single supplier, while the good material in the original archive had come from various other suppliers. At this point, it was no surprise that this single supplier was the same one whose good materials were now being improperly rejected by the analyzer. As you can see, although we thought we had developed a great calibration to distinguish acceptable from unacceptable feedstock, we had, instead, developed a calibration that was extremely good at determining which feedstock was provided by that one particular supplier, regardless of the acceptability/rejectability of the feedstock!

As unpleasant as the whole episode was, it could have been much worse. The process was running with mass inputs costing nearly $100,000 per day. If, instead of wrongly rejecting good materials, the system had wrongly accepted bad materials, the losses due to production of worthless scrap would have been considerable indeed!

So here is a case where the data had plenty of information, but the information in the data was not correlated to the property which was being predicted. While there is no way to completely protect yourself from this type of problem, an active and agressive cynicism certainly doesn't hurt. Trust nothing—question everything!

Examples of Data

An exhaustive list of all possible types of data suitable for chemometric treatment together with all possible types of predictions made from the data would fill a large chapter in this book. Table 1 contains a brief list of some of these. Table 1 is like a Chinese menu—selections from the first column can be freely paired with selections from the second column in almost any permutation. Notice that many data types may serve either as the measured data or the predicted property, depending upon the particular application.

We tend to think that the data we start with is usually some type of instrumental measurement like a spectrum or a chromatogram, and that we are usually trying to predict the concentrations of various components, or the thickness of various layers in a sample. But, as illustrated in Table 1, we can use almost any sort of data to predict almost anything, as long as there is some relationship between the information in the data and the property which we are trying to predict. For example we might start with measurements of pH, temperatures, stirring rates, and reaction times, for a process and use these data to predict the tensile strength, or hardness of the resulting product. Or we might

MEASUREMENT	PREDICTION
Spectrum	Concentrations
Chromatogram	Purity
Interferogram	Physical Properties
Physical Properties	Source or Origin
Temperature	Accept/Reject
Identity	Reaction End Point
Pressure	Chemical Properties
Surface Acoustic Wave Response	Concentrations
Concentrations	Source or Origin
Molecular Weights	Rheology
Structure	Biological Activity
Stability	Structure
pH	Temperature
Flow	Age

Table 1. Some types of data and predicted parameters

measure the viscoscity, vapor pressure, and trace element concentrations of a material and use them to identify the manufacturer of the material, or to classify the material as acceptable or unacceptable for a particular application.

When considering potential applications for these techniques, there is no reason to restrict our thinking as to which particular types of data we might use or which particular kinds of properties we could predict. Reflecting the generality of these techniques, mathematicians usually call the measured data the *independent variables*, or the *x-data*, or the *x-block data*. Similarly, the properties we are trying to predict are usually called the *dependent variables*, the *y-data*, or the *y-block data*. Taken together, the set of corresponding x and y data measured from a single sample is called an *object*. While this system of nomenclature is precise, and preserves the concept of the generality of the methods, many people find that this nomenclature tends to "get between" them and their data. It can be a burdensome distraction when you constantly have to remember which is the x-data and which is the y-data. For this reason, throughout the remainder of the book, we will adopt the vocabulary of spectroscopy to discuss our data. We will imagine that we are measuring an absorbance spectrum for each of our samples and that we want to predict the concentrations of the constituents in the samples. But please remember, we are adopting this vocabulary merely for convenience. The techniques themselves can be applied for myriad purposes other than quantitative spectroscopic analysis.

Data Organization

As we will soon see, the nature of the work makes it extremely convenient to organize our data into matrices. (If you are not familiar with data matrices, please see the explanation of matrices in Appendix A before continuing.) In particular, it is useful to organize the dependent and independent variables into separate matrices. In the case of spectroscopy, if we measure the absorbance spectra of a number of samples of known composition, we assemble all of these spectra into one matrix which we will call the *absorbance matrix*. We also assemble all of the concentration values for the sample's components into a separate matrix called the *concentration matrix*. For those who are keeping score, the absorbance matrix contains the independent variables (also known as the x-data or the x-block), and the concentration matrix contains the dependent variables (also called the y-data or the y-block).

The first thing we have to decide is whether these matrices should be organized *column-wise* or *row-wise*. The spectrum of a single sample consists of the individual absorbance values for each wavelength at which the sample was measured. Should we place this set of absorbance values into the absorbance matrix so that they comprise a column in the matrix, or should we place them into the absorbance matrix so that they comprise a row? We have to make the same decision for the concentration matrix. Should the concentration values of the components of each sample be placed into the concentration matrix as a row or as a column in the matrix? The decision is totally arbitrary, because we can formulate the various mathematical operations for either row-wise or column-wise data organization. But we do have to choose one or the other. Since Murphy established his laws long before chemometricians came on the scene, it should be no surprise that *both* conventions are commonly employed throughout the literature!

Generally, the Multiple Linear Regression (MLR) techniques and the Factor-Based technique known as Principal Component Regression (PCR) employ data that is organized as matrices of column vectors, while the Factor-Based technique known as Partial Least-Squares (PLS) employs data that is organized as matrices of row vectors. The conflicting conventions are simply the result of historical accident. Some of the first MLR work was pioneered by spectroscopists doing quantitative work with Beer's law. The way spectroscopists write Beer's law is consistent with column-wise organization of the data matrices. When these pioneers began exploring PCR techniques, they retained the column-wise organization. The theory and practice of PLS was developed around work in other fields of science. The problems being addressed in those fields were more conveniently handled with data that was organized as matrices of row vectors. When chemometricians began to adopt the PLS techniques, they also adopted the row-wise convention. But, by that point in time, the column-wise convention for MLR and PCR was well established. So we are stuck with a dual set of conventions. To complicate things even further most of the MLR and PCR work in the field of near infrared spectroscopy (NIR) employs the row-wise convention.

Column-Wise Data Organization for MLR and PCR Data

Absorbance Matrix

Using column-wise organization, an absorbance matrix holds the spectral data. Each spectrum is placed into the absorbance matrix as a column vector:

$$
\begin{matrix}
A_{11} & A_{12} & A_{13} & \dots & A_{1s} \\
A_{21} & A_{22} & A_{23} & \dots & A_{2s} \\
A_{31} & A_{32} & A_{33} & \dots & A_{3s} \\
A_{41} & A_{42} & A_{43} & \dots & A_{4s} \\
\dots & \dots & \dots & \dots & \dots \\
A_{w1} & A_{w2} & A_{w3} & \dots & A_{ws}
\end{matrix}
\qquad [1]
$$

where A_{ws} is the absorbance at the wth wavelength for sample **s**. If we were to measure the spectra of 30 samples at 15 different wavelengths, each spectrum would be held in a column vector containing 15 absorbance values. These 30 column vectors would be assembled into an absorbance matrix which would be 15 X 30 in size (15 rows, 30 columns). Another way to visualize the data organization is to represent each column vector containing each absorbance spectrum as a line drawing —

either drawn so,   or so,   or so:

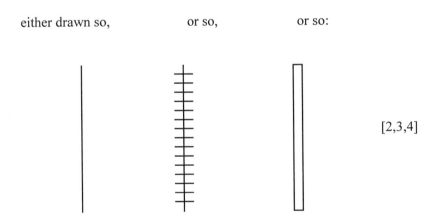

[2,3,4]

The corresponding absorbance matrix (shown with only 3 spectra) would be represented —

either drawn so,   or so,   or so,   or so:

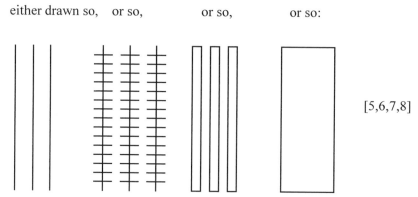

[5,6,7,8]

Concentration Matrix

Similarly, a concentration matrix holds the concentration data. The concentrations of the components for each sample are placed into the concentration matrix as a column vector:

$$
\begin{matrix}
C_{11} & C_{12} & C_{13} & ... & C_{1s} \\
C_{21} & C_{22} & C_{23} & ... & C_{2s} \\
... & ... & ... & & ... \\
C_{c1} & C_{c2} & C_{c3} & ... & C_{cs}
\end{matrix}
\qquad [9]
$$

Where C_{cs} is the concentration of the cth component of sample s. Suppose we were measuring the concentrations of 4 components in each of the 30 samples, above. The concentrations for each sample would be held in a column vector containing 4 concentration values. These 30 column vectors would be assembled into a concentration matrix which would be 4 X 30 in size (4 rows, 30 columns).

Taken together, the absorbance matrix and the concentration matrix comprise a data set. It is essential that the columns of the absorbance and concentration matrices correspond to the same mixtures. In other words, the sth column of the absorbance matrix *must* contain the spectrum of the sample whose component concentrations are contained in the sth column of the concentration matrix. A data set for a single sample, would comprise an absorbance matrix with a single column containing the spectrum of that sample together with a corresponding concentration matrix with a single column containing the concentrations of the components of that sample. As explained earlier, such a data set comprising a single sample is often called an *object*.

A data matrix with column-wise organization is easily converted to row-wise organization by taking its matrix transpose, and vice versa. If you are not familiar with the matrix transpose operation, please refer to the discussion in Appendix A.

Row-Wise Data Organization for PLS Data

Absorbance Matrix

Using row-wise organization, an absorbance matrix holds the spectral data. Each spectrum is placed into the absorbance matrix as a row vector:

$$\begin{array}{cccccc}
A_{11} & A_{12} & A_{13} & A_{14} & \ldots & A_{1w} \\
A_{21} & A_{22} & A_{23} & A_{24} & \ldots & A_{2w} \\
A_{31} & A_{32} & A_{33} & A_{34} & \ldots & A_{3w} \\
\ldots & \ldots & \ldots & \ldots & \ldots & \ldots \\
A_{s1} & A_{s2} & A_{s3} & A_{s4} & \ldots & A_{sw}
\end{array} \qquad [10]$$

Where A_{sw} is the absorbance for sample s at the wth wavelength. If we were to measure the spectra of 30 samples at 15 different wavelengths, each spectrum would be held in a row vector containing 15 absorbance values. These 30 row vectors would be assembled into an absorbance matrix which would be 30 X 15 in size (30 rows, 15 columns).

Another way to visualize the data organization is to represent the row vector containing the absorbance spectrum as a line drawing —

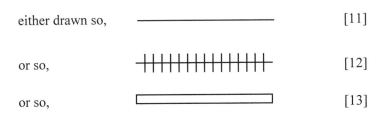

either drawn so,   [11]

or so,   [12]

or so,   [13]

The corresponding absorbance matrix (shown with 3 spectra) would be represented —

either drawn so,   [14]

or so,   [15]

or so,   [16]

or so: [17]

Concentration Matrix

Similarly, a concentration matrix holds the concentration data. The concentrations of the components for each sample are placed into the concentration matrix as a row vector:

$$
\begin{matrix}
C_{11} & C_{12} & \cdots & C_{1c} \\
C_{21} & C_{22} & \cdots & C_{2c} \\
C_{31} & C_{32} & \cdots & C_{3c} \\
C_{41} & C_{42} & \cdots & C_{4c} \\
\cdots & \cdots & \cdots & \cdots \\
C_{s1} & C_{s2} & \cdots & C_{sc}
\end{matrix}
\qquad [18]
$$

Where C_{sc} is the concentration for sample s of the cth component. Suppose we were measuring the concentrations of 4 components in each of the 30 samples, above. The concentrations for each sample would be held in a row vector containing 4 concentration values. These 30 row vectors would be assembled into a concentration matrix which would be 30 X 4 in size (30 rows, 4 columns).

Taken together, the absorbance matrix and the concentration matrix comprise a data set. It is essential that the rows of the absorbance and concentration matrices correspond to the same mixtures. In other words, the sth row of the absorbance matrix *must* contain the spectrum of the sample whose component concentrations are contained in the sth row of the concentration matrix. A data set for a single sample, would comprise an absorbance matrix with a single row containing the spectrum of that sample together with a corresponding concentration matrix with a single row containing the concentrations of the components of that sample. As explained earlier, such a data set comprising a single sample is often called an *object*.

A data matrix with row-wise organization is easily converted to column-wise organization by taking its matrix transpose, and vice versa. If you are not familiar with the matrix transpose operation, please refer to the discussion in Appendix A.

Data Sets

We have seen that data matrices are organized into pairs; each absorbance matrix is paired with its corresponding concentration matrix. The pair of matrices comprise a *data set*. Data sets have different names depending on their origin and purpose.

Training Set

A data set containing measurements on a set of known samples and used to develop a calibration is called a *training set*. The known samples are sometimes called the *calibration samples*. A training set consists of an absorbance matrix containing spectra that are measured as carefully as possible and a concentration matrix containing concentration values determined by a reliable, independent referee method.

The data in the training set are used to derive the calibration which we use on the spectra of unknown samples (i.e. samples of unknown composition) to predict the concentrations in those samples. In order for the calibration to be valid, the data in the training set which is used to find the calibration must meet certain requirements. Basically, the training set must contain data which, as a group, are representative, *in all ways*, of the unknown samples on which the analysis will be used. A statistician would express this requirement by saying, "The training set must be a statistically valid sample of the population comprised of all unknowns on which the calibration will be used." Additionally, because we will be using multivariate techniques, it is very important that the samples in the training set are all mutually independent.

In practical terms, this means that training sets should:

1. Contain all expected components
2. Span the concentration ranges of interest
3. Span the conditions of interest
4. Contain mutually independent samples

Let's review these items one at a time.

Contain All Expected Components

This requirement is pretty easy to accept. It makes sense that, if we are going to generate a calibration, we must construct a training set that exhibits all the forms of variation that we expect to encounter in the unknown samples. We certainly would not expect a calibration to produce accurate results if an unknown sample contained a spectral peak that was never present in any of the calibration samples.

However, many find it harder to accept that "components" must be understood in the broadest sense. "Components" in this context does not refer solely to a sample's constituents. "Components" must be understood to be synonymous with "sources of variation." We might not normally think of instrument drift as a "component." But a change in the measured spectrum due to drift in the instrument is indistinguishable from a change in the measured spectrum due to the presence of an additional component in the sample. Thus, instrument drift is, indeed, a "component." We might not normally think that replacing a sample cell would represent the addition of a new component. But subtle differences in the construction and alignment of the new sample cell might add artifacts to the specturm that could compromise the accuracy of a calibration. Similarly the differences in technique between two instrument operators could also cause problems.

Span the Concentration Ranges of Interest

This requirement also makes good sense. A calibration is nothing more than a mathematical model that relates the behavior of the measureable data to the behavior of that which we wish to predict. We construct a calibration by finding the best representation of the *fit* between the measured data and the predicted parameters. It is not surprising that the performance of a calibration can deteriorate rapidly if we use the calibration to extrapolate predictions for mixtures that lie further and further outside the concentration ranges of the original calibration samples.

However, it is not obvious that when we work with multivariate data, our training set must span the concentration ranges of interest in a multivariate (as opposed to univariate) way. It is not sufficient to create a series of samples where each component is varied individually while all other components are held constant. Our training set must contain data on samples where all of the various components (remember to understand "components" in the broadest sense) vary simultaneously *and* independently. More about this shortly.

Span the Conditions of Interest

This requirement is just an additional broadening of the meaning of "components." To the extent that variations in temperature, pH, pressure, humidity, environmental factors, etc., can cause variations in the spectra we measure, such variations must be represented in the training set data.

Mutual Independence

Of all the requirements, mutual independence is sometimes the most difficult one to appreciate. Part of the problem is that the preparation of mutually independent samples runs somewhat countrary to one of the basic techniques for sample preparation which we have learned, namely serial dilution or addition. Nearly everyone who has been through a lab course has had to prepare a series of calibration samples by first preparing a stock solution, and then using that to prepare a series of successively more dilute solutions which are then used as standards. While these standards might be perfectly suitable for the generation of a simple, univariate calibration, they are entirely unsuitable for calibrations based on multivariate techniques. The problem is that the relative concentrations of the various components in the solution are not varying. Even worse, the relative errors among the concentrations of the various components are not varying. The only varying sources of error are the overall dilution error, and the instrumental noise.

Validation Set

It is highly desireable to assemble an additional data set containing independent measurements on samples that are independent from the samples used to create the training set. This data set is not used to develop the calibration. Instead, it is held in reserve so that it can be used to evaluate the calibration's performance. Samples held in reserve this way are known as *validation samples* and the pair of absorbance and concentration matrices holding these data is called a *validation set*.

The data in the validation set are used to challenge the calibration. We treat the validation samples as if they are unknowns. We use the calibration developed with the training set to predict (or estimate) the concentrations of the components in the validation samples. We then compare these predicted concentrations to the actual concentrations as determined by an independent referee method (these are also called the *expected* concentrations). In this way, we can assess the expected performance of the calibration on actual unknowns. To the extent that the validation samples are a good representation of all the unknown samples we will encounter, this validation step will provide a reliable estimate of the calibration's performance on the unknowns. But if we encounter unknowns that are significantly different from the validation samples, we are likely to be surprised by the actual performance of the calibration (and such surprises are seldom pleasant).

Unknown Set

When we measure the spectrum of an unknown sample, we assemble it into an absorbance matrix. If we are measuring a single unknown sample, our unknown absorbance matrix will have only one column (for MLR or PCR) or one row (for PLS). If we measure the spectra of a number of unknown samples, we can assemble them together into a single unknown absorbance matrix just as we assemble training or validation spectra.

Of course, we cannot assemble a corresponding unknown concentration matrix because we do not know the concentrations of the components in the unknown sample. Instead, we use the calibration we have developed to calculate a *result matrix* which contains the predicted concentrations of the components in the unknown(s). The result matrix will be organized just like the concentration matrix in a training or validation data set. If our unknown absorbance matrix contained a single spectrum, the result matrix will contain a single column (for MLR or PCR) or row (for PLS). Each entry in the column (or row) will be the concentration of each component in the unknown sample. If our unknown absorbance matrix contained multiple spectra, the result matrix will contain one column (for MLR or PCR) or one row (for PLS) of concentration values for the sample whose spectrum is contained in the corresponding column or row in the unknown absorbance matrix. The absorbance matrix containing the unknown spectra together with the corresponding result matrix containing the predicted concentrations for the unknowns comprise an *unknown set*.

Basic Approach

The flow chart in Figure 1 illustrates the basic approach for developing calibrations and placing them successfully into service. While this approach is simple and straightforward, putting it into practice is not always easy. The concepts summarized in Figure 1 represent the most important information in this entire book — to ignore them is to invite disaster. Accordingly, we will discuss each step of the process in some detail.

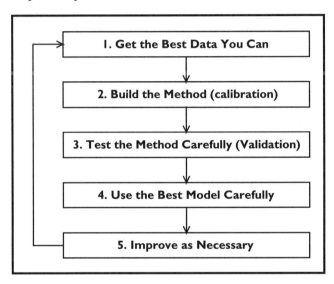

Figure 1. Flow chart for developing and using calibrations.

Get the Best Data You Can

This first step is often the most difficult step of all. Obviously, it makes sense to work with the best data you can get your hands on. What is not so obvious is the definition of *best*. To arrive at an appropriate definition for a given application, we must balance many factors, among them:

1. Number of samples for the training set
2. Accuracy of the concentration values for the training set
3. Number of samples in the validation set (if any)
4. Accuracy of the concentration values for the validation set
5. Noise level in the spectra

We can see that the cost of developing and maintaining a calibration will depend strongly on how we choose among these factors. Making the right choices is particularly difficult because there is no single set of choices that is appropriate for all applications. The best compromise among cost and effort put into the calibration vs. the resulting analytical performance and robustness must be determined on a case by case basis.

The situation can be complicated even further if the managers responsible for allocating resources to the project have an unrealistic idea of the resources which must be committed in order to successfuly develop and deploy a calibration. Unfortunately, many managers have been "oversold" on chemometrics, coming to believe that these techniques represent a type of "black magic" which can easily produce pristine calibrations that will 1) perform properly the first day they are placed in service and, 2) without further attention, continue to perform properly, in perpetuity. This illusion has been reinforced by the availablity of powerful software that will happily produce "calibrations" at the push of a button using any data we care to feed it. While everyone understands the concept of "garbage in—garbage out", many have come to believe that this rule is suspended when chemometrics are put into play.

If your managers fit this description, then forget about developing any chemometric calibrations without first completing an absolutly essential initial task: The Education of Your Managers. If your managers do not have realistic expections of the capabilities and limitations of chemometric calibrations, and/or if they do not provide the informed commitment of adequate resources, your project is guaranteed to end in grief. Educating your managers can be the most difficult and the most important step in successfully applying these techniques.

Rules of Thumb

It may be overly optimistic to assume that we can freely decide how many samples to work with and how accurately we will measure their concentrations. Often there are a very limited number of calibration samples available and/or the accuracy of the samples' concentration values is miserably poor. Nonetheless, it is important to understand, from the outset, what the tradeoffs are, and what would normally be considered an adequate number of samples and adequate accuracy for their concentration values.

This isn't to say that it is impossible to develop a calibration with fewer and/or poorer samples than are normally desireable. Even with a limited number of poor samples, we might be able to "bootstrap" a calibration with a little luck, a lot of labor, and a healthy dose of skepticism.

The rules of thumb discussed below have served this author well over the years. Depending on the nature of your work and data, your experiences may lead you to modify these rules to suit the particulars of your applications. But they should give you a good place to start.

Training Set Concentration Accuracy

All of these chemometric techniques have one thing in common. The analytical performance of a calibration deteriorates rapidly as the accuracy of the concentration values for the training set samples deteriorates. What's more, any advantages that the factor based techniques might offer over the ordinary multiple linear regressions disappear rapidly as the errors in the training set concentration values increase. In other words, improvements in the accuracy of a training set's concentration values can result in major improvements in the analytical performance of the calibration developed from that training set.

In practical terms, we can usually develop satisfactory calibrations with training set concentrations, as determined by some referee method, that are accurate to ±5% mean relative error. Fortunately, when working with typical industrial applications and within a reasonable budget, it is usually possible to achieve at least this level of accuracy. But there is no need to stop there. We will usually realize significant benefits such as improved analytical accuracy, robustness, and ease of calibration if we can reduce the errors in the training set concentrations to   ±2% or ±3%. The benefits are such that it is usually worthwhile to shoot for this level of accuracy whenever it can be reasonably achieved.

Going in the other direction, as the errors in the training set concentrations climb above ±5%, life quickly becomes umpleasant. In general, it can be difficult to achieve useable results when the concentration errors rise above ±10%

Number of Calibration Samples in the Training Set

There are three rules of thumb to guide us in selecting the number of calibration samples we should include in a training set. They are all based on the number of components in the system with which we are working. Remember that components should be understood in the widest sense as "independent sources of significant variation in the data." For example, a system with 3 constituents that is measured over a range of temperatures would have at least 4 components: the 3 constituents plus temperature.

The **Rule of 3** is the minimum number of samples we should normally attempt to work with. It says, simply, "Use 3 times the number of samples as there are components." While it is possible to develop calibrations with fewer samples, it is difficult to get acceptable calibrations that way. If we were working with the above example of a 4-component system, we would expect to need at least 12 samples in our training set. While the Rule of 3 gives us the minimum number of samples we should normally attempt to use, it is not a comfortable minimum. We would normally employ the Rule of 3 only when doing preliminary or exploratory work.

The **Rule of 5** is a better guide for the minimum number of samples to use. Using 5 times the number of samples as there are components allows us enough samples to reasonably represent all possible combinations of concentrations values for a 3-component system. However, as the number of components in the system increases, the number of samples we should have increases geometrically. Thus, the Rule of 5 is not a comfortable guide for systems with large numbers of components.

The **Rule of 10** is better still. If we use 10 times the number of samples as there are components, we will usually be able to create a solid calibration for typical applications. Employing the Rule of 10 will quickly sensitize us to the need we discussed earlier of Educating the Managers. Many managers will balk at the time and money required to assemble 40 calibration samples (considering the example, above, where temperature variations act like a 4th component) in order to generate a calibration for a "simple" 3 constituent system. They would consider 40 samples to be overkill. But, if we want to reap the benefits that these techniques can offer us, 40 samples is not overkill in any sense of the word.

You might have followed some of the recent work involving the use of chemometrics to predict the octane of gasoline from its near infrared (NIR) spectrum. Gasoline is a rather complex mixture with not dozens, but hundreds of constituents. The complexity is increased even further when you consider that a practical calibration has to work on gasoline produced at multiple refineries and blended differently at different times of the year. During some of the early discussion of this application it was postulated that, due to the complexity of the system, several hundred samples might be needed in the training set. (Notice the consistency with the Rule of 3 or the Rule of 5.) The time and cost involved in assembling measurements on several hundred samples was a bit discouraging. But, since this is an application with tremendous payback potential, several companies proceeded, nonetheless, to develop calibrations. As it turns out, the methods that have been successfully deployed after many years of development are based on training sets containing several *thousand* calibration samples. Even considering the number of components in gasoline, the Rule of 10 did not overstate the number of samples that would be necessary.

We must often compromise between the number of samples in the training set and the accuracy of the concentration values for those samples. This is because the additional time and money required for a more accurate referee method for determining the concentrations must often be offset by working with fewer samples. The more we know about the particulars of an application, the easier it would be for us to strike an informed compromise. But often, we don't know as much as we would like.

Generally, if the accuracy and precision of a calibration is an overriding concern, it is often a good bet to back down from the Rule of 10 and compromise on the Rule of 5 if we can thereby gain at least a factor of 3 improvement in the accuracy of the training set concentrations. On the other hand, if a calibration's long term reliability and robustness is more important than absolute accuracy or precision, then it would generally be better to stay with the Rule of 10 and forego the improved concentration accuracy.

Build the Method (calibration)

Generating the calibration is often the easiest step in the whole process thanks to the widespread availability of powerful, inexpensive computers and capable software. This step is often as easy as moving the data into a computer, making a few simple (but well informed!) choices, and pushing a few keys on the keyboard. This step will be covered in the remaining chapters of this book.

Test the Method Carefully (validation)

The best protection we have against placing an inadequate calibration into service is to challenge the calibration as agressively as we can with as many validation samples as possible. We do this to uncover any weaknesses the calibration might have and to help us understand the calibration's limitations. We pretend that the validation samples are unknowns. We use the calibration that we developed with the training set to predict the concentrations of the validation samples. We then compare these predicted concentrations to the known or expected concentrations for these samples. The error between the predicted concentrations vs. the expected values is indicative of the error we could expect when we use the calibration to analyze actual unknown samples.

This is another aspect of the process about which managers often require some education. After spending so much time, effort, and money developing a calibration, many managers are tempted to rush it into service without adequate validation. The best way to counter this tendency is to patiently explain that we do not have the ability to choose *whether or not* we will validate a calibration. We only get to choose *where* we will validate it. We can either choose to validate the calibration at development time, under controlled conditions, or we can choose to validate the method by placing it into service and observing whether or not it is working properly— while hoping for the best. Obviously, if we place a calibration into service without first adequately testing it, we expose ourselves to the risk of expensive losses should the method prove inadequate for the application.

Ideally, we validate a calibration with a great number of validation samples. Validation samples are samples that were not included in the training set. They should be as representative as possible of all of the unknown samples which the calibration is expected to successfully analyze. The more validation samples we use, and the better they represent all the different kinds of unknowns we might see, the greater the liklihood that we will catch a situation or a sample where the calibration will fail. Conversely, the fewer validation samples we use, the more likely we are to encounter an unpleasant surprise when we put the calibration into service— especially if these relatively few validation samples we are "easy cases" with few anomalies.

Whenever possible, we would prefer that the concentration values we have for the validation samples are as accurate as the training set concentration values. Stated another way, we would like to have enough calibration samples to construct the training set plus some additional samples that we can hold in reserve for use as validation samples. Remember, validation samples, by definition, cannot be used in the training set. (However, *after* the validation process is *completed*, we could then decide to incorporate the validation samples into the training set and recalculate the calibration on this larger data set. This will usually improve the calibration's accuracy and robustness. We would not want to use the validation samples this way if the accuracy of their concentrations is significantly poorer than the accuracy of the training set concentrations.)

We often cannot afford to assemble large numbers of validations samples with concentrations as accurate as the training set concentrations. But since the validation samples are used to *test* the calibration rather than *produce* the calibration, errors in validation sample concentrations do not have the same detrimental impact as errors in the training set concentrations. Validation set concentration errors cannot affect the calibration model. They can only make it more difficult to understand how well or poorly the calibration is working. The effect of validation concentration errors can be averaged out by using a large number of validation samples.

Rules of Thumb

Number of Calibration Samples in the Validation Set

Generally speaking, the more validation samples the better. It is nice to have at least as many samples in the validation set as were needed in the training set. It is even better to have considerably more validation samples than calibration samples.

Validation Set Concentration Accuracy

Ideally, the validation concentrations should be as accurate as the training concentrations. However, validation samples with poorer concentration accuracy are still useful. In general, we would prefer that validation concentrations would not have errors greater than ±5%. Samples with concentrations errors of around ±10% can still be useful. Finally, validation samples with concentration errors approaching ±20% are better than no validation samples at all.

Validation Without Validation Samples

Sometimes it is just not feasible to assemble any validation samples. In such cases there are still other tests, such as cross-validation, which can help us do a certain amount of validation of a calibration. However, these tests do not provide the level of information nor the level of confidence that we should have before placing a calibration into service. More about this later.

Use the Best Model Carefully

After a calibration is created and properly validated, it is ready to be placed into service. But our work doesn't end here. If we simply release the method and walk away from it, we are asking for trouble. The model must be used *carefully*.

There are many things that go into the concept of "carefully." For these purposes, "carefully" means "with an appropriate level of cynicism." "Carefully" also means that proper procedures must be put into place, and that the people who rely on the results of the calibration must be properly trained to use the calibration.

We have said that every time the calibration analyzes a new unknown sample, this amounts to an additional validation test of the calibration. It can be a major mistake to believe that, just because a calibration worked well when it was being developed, it will continue to produce reliable results from that point on. When we discussed the requirements for a training set, we said that collection of samples in the training set must, as a group, be representative *in all ways* of the unknowns that will be analyzed by the calibration. If this condition is not met, then the calibration is invalid and cannot be expected to produce reliable results. Any change in the process, the instrument, or the measurement procedure which introduces changes into the data measured on an unknown will violate this condition and invalidate the method! If this occurs, the concentration values that the calibration predicts for unknown samples are completely unreliable! We must therefore have a plan and procedures in place that will insure that we are alerted if such a condition should arise.

Auditing the Calibration

The best protection against this potential for unreliable results is to collect samples at appropriate intervals, use a suitable referee method to independently determine the concentrations of these samples, and compare the referee concentrations to the concentrations predicted by the calibration. In other words, we institute an on-going program of validation as long as the method is in service. These validation samples are sometimes called *audit samples* and this on-going validation is sometimes called *auditing* the calibration. What would constitute an appropriate time interval for the audit depends very much on the nature of the process, the difficulty of the analysis, and the potential for changes. After first putting the method into service, we might take audit samples every hour. As we gain confidence in the method, we might reduce the frequency to once or twice a shift, then to once or twice a day, and so on.

Training

It is essential that those involved with the operation of the process, and the calibration as well as those who are relying on the results of the calibration have a basic understanding of the vulnerability of the calibration to unexpected changes. The maintenance people and instrument technicians must understand that if they change a lamp or clean a sample system, the analyzer might start producing wrong answers. The process engineers must understand that a change in operating conditions or feedstock can totally confound even the best calibration. The plant manager must understand the need for periodic audit samples, and the need to document what otherwise might seem to be inconsequential details.

Procedures

Finally, when a calibration is put into service, it is important that proper procedures are simultaneously put into place throughout the organization. These procedures involve not only actions that must occur, such as collection and analysis of audit samples, but also communication that must take place. For example, if the purchasing department were considering changing the supplier of a feedstock, they might consult with the chemical engineer or the manufacturing engineer responsible for the process in question, but it is unlikely that any of these people would realize the importance of consulting with you, the person responsible for developing and installing the analyzer using a chemometric calibration. Yet, a change in feedstock could totally cripple the calibration you developed. Similarly, it is seldom routine practice to notify the analytical chemist responsible for an analyzer if there is a change in operating or maintenance people. Yet, the performance of an analyzer can be sensitive to differences in sample preparation technique, sample system maintenace and cleaning, etc. So it might be necessary to increase the frequency of audit samples if new people are trained on an analyzer. Every application will involve different particulars. It is important that you do not develop and install a calibration in a vacuum. Consider all of the operational issues that might impact on the reliability of the analysis and design your procedures and train your people accordingly.

Improve as Necessary

An effective auditing plan allows us to identify and address any difficiencies in the calibration, and/or to improve the calibration over the course of time. At the very least, so long as the accuracy of the concentration values determined by the referee method is at least as good as the accuracy of the original calibration samples, we can add the audit samples to the training set and recalculate the calibration. As we incorporate more and more samples into the training set, we capture more and more sources of variation in the data. This should make our calibration more and more robust, and it will often improve the accuracy as well. In general, as instruments and sample systems age, and as processes change, we will usually see a gradual, but steady deterioration in the performance of the initial calibration. Periodic updating of the training set, can prevent the deterioration.

Incremental updating of the calibration, while it is very useful, is not sufficient in every case. For example, if there is a significant change in the application, such as a change in trace contaminants due to a change in feedstock supplier, we might have to discard the original calibration and build a new one from scratch.

Creating Some Data

It is time to create some data to play with. By creating the data ourselves, we will know exactly what its properties are. We will subject these data to each of the chemometric techniques so that we may observe and discuss the results. We will be able to translate our detailed *a priori* knowledge of the data into a detailed understanding of how the different techniques function. In this way, we will learn the strengths and weaknesses of the various methods and how to use them correctly.

As discussed in the first chapter, it is possible to use almost any kind of data to predict almost any type of property. But to keep things simple, we will continue using the vocabulary of spectroscopy. Accordingly, we will call the data we create *absorbance spectra*, or simply *spectra*, and we will call the property we are trying to predict *concentration*.

In order to make this exercise as useful and as interesting as possible, we will take steps to insure that our synthetic data are suitably realistic. We will include difficult spectral interferences, and we will add levels of noise and other artifacts that might be encountered in a typical, industrial application.

Synthetic Data Sets

As we will soon see, the most difficult part of working with these techniques is keeping track of the large amounts of data that are usually involved. We will be constructing a number of different data sets, and we will find it necessary to constantly review which data set we are working with at any particular time. A data "crib sheet" has been included on the inside back cover of this book to help us with this task.

To (hopefully) help keep things simple, we will organize all of our data into column-wise matrices. Later on, when we explore Partial Least-Squares (PLS), we will have to remember that the PLS convention expects data to be organized row-wise. This isn't a great problem since one convention is merely the matrix transpose of the other. Nonetheless, it is one more thing we have to remember.

Our data will simulate spectra collected on mixtures that contain 4 different components dissolved in a spectrally inactive solvent. We will suppose that we have measured the concentrations of 3 of the components with referee methods. The 4th component will be present in varying amounts in all of the samples, but we will not have access to any information about the concentrations of the 4th component.

We will organize our data into training sets and validation sets. The training sets will be used to develop the various calibrations, and the validation sets will be used to evaluate how well the calibrations perform.

Training Set Design

A calibration can only be as good as the training set which is used to generate it. We must insure that the training set accurately represents all of the unknowns that the calibration is expected to analyze. In other words, the training set must be a statistically valid sample of the population comprising all unknown samples on which the calibration will be used.

There is an entire discipline of Experimental Design that is devoted to the art and science of detemining the what should be in a training set. A detailed exploration of the Design of Experiments (DOE, or experimental design) is beyond the scope of this book. Please consult the bibliography for publications that treat this topic in more detail.

The first thing we must understand is that these chemometric techniques do not usually work well when they are used to analyze samples by extrapolation. This is true regardless of how linear our system might be. To prevent extrapolation, the concentrations of the components in our training set samples must span the full range of concentrations that will be present in the unknowns. The next thing we must understand is that we are working with *multivariate* systems. In other words, we are working with samples whose component concentrations, in general, vary independently of one another. This means that, when we talk about spanning the full range of concentrations, we have to understand the concept of spanning in a multivariate way. Finally, we must understand how to visualize and think about multivariate data.

Figure 2 is a multivariate plot of some multivariate data. We have plotted the component concentrations of several samples. Each sample contains a different combination of concentrations of 3 components. For each sample, the concentration of the first component is plotted along the x-axis, the concentration of the second component is plotted along the y-axis, and the concentration of the third component is plotted along the z-axis. The concentration of each component will vary from some minimum value to some maximum value. In this example, we have arbitrarily used zero as the minimum value for each component concentration and unity for the maximum value. In the real world, each component could have a different minimum value and a different maximum value than all of the other components. Also, the minimum value need not be zero and the maximum value need not be unity.

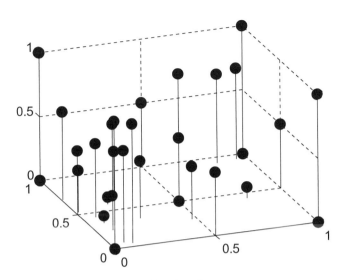

Figure 2. Multivariate view of multivariate data.

When we plot the sample concentrations in this way, we begin to see that each sample with a unique combination of component concentrations occupies a unique point in this *concentration space*. (Since this is the concentration space of a training set, it sometimes called the *calibration space*.) If we want to construct a training set that spans this concentration space, we can see that we must do it in the multivariate sense by including samples that, taken as a set, will occupy all the relevant portions of the concentration space.

Figure 3 is an example of the *wrong* way to span a concentration space. It is a plot of a training set constructed for a 3-component system. The problem with this training set is that, while a large number of samples are included, and the concentration of each component is varied through the full range of expected concentration values, every sample in the set contains only a single component. So, even though the samples span the full range of concentrations, they do not span the full range of the possible *combinations* of the concentrations. At best, we have spanned that portion of the concentration space indicated by the shaded volume. But since all of the calibration samples lie along only 3 edges of this 6-edged shaded volume, the training set does not even span the shaded volume properly. As a consequense, if we generate a calibration with this training set and use it to predict the concentrations of the sample "X" plotted in Figure 3, the calibration will actually be doing an extrapolation. This is true even though the concentrations of the individual components in sample X do not exceed the

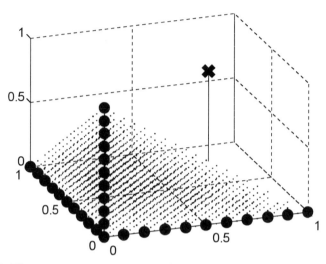

Figure 3. The wrong way to span a multivariate data space.

concentrations of those components in the training set samples. The problem is that sample X lies outside the region of the calibration space spanned by the samples in the training set. One common feature of all of these chemometric techniques is that they generally perform poorly when they are used to extrapolate in this fashion.There are three main ways to construct a proper multivariate training set:

1. Structured
2. Random
3. Manually

Structured Training Sets

The structured approach uses one or more systematic schemes to span the calibration space. Figure 4 illustrates, for a 3-component system, one of the most commonly employed structured designs. It is usually known as a *full-factorial* design. It uses the minimum, maximum, and (optionally) the mean concentration values for each component. A sample set is constructed by assembling samples containing all possible combinations of these values. When the mean concentration values are not included, this approach generates a training set that fully spans the concentration space with the fewest possible samples. We see that this approach gives us a calibration sample at every vertex of the calibration of the calibration space. When the mean concentration values are used we also have a sample in the center of each face of the calibration space, one sample in the center of each edge of the calibration space, and one sample in the center of the space.

For our purposes, we would generally prefer to include the mean concentrations for two reasons. First of all, we usually want to have more samples in the training set than we would have if we leave the mean concentration values out of the factorial design. Secondly, if we leave out the mean concentration values, we only get samples at the vertices of the calibration space. If our spectra change in a perfectly linear fashion with the variations in concentration, this would not be a concern. However, if we only have samples at the vertices of the calibration space, we will not have any way of detecting the presence of nonlinearities nor will the calibration be able to make any attempt to compensate for them. When we generate the calibration with such a training set, the calculations we employ will minimize the errors only for these samples at the vertices since those are the only samples there are. In the presence of nonlinearities, this could result in an undesireable increase in the errors for the central regions of the space. This problem can be severe if our data contain significant nonlinearities. By including the samples with the mean concentration values in the training set, we help insure that calibration errors are

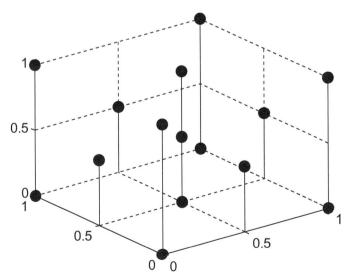

Figure 4. Concentrations values of a structured training set.

not minimized at the vertices at the expense of the central regions. The bottom line is that calibrations based on training sets that include the mean concentrations tend to produce better predictions on typical unknowns than calibrations based on training sets that exclude the mean concentrations.

Random Training Sets

The random approach involves randomly selecting samples throughout the calibration space. It is important that we use a method of random selection that does not create an underlying correlation among the concentrations of the components. As long as we observe that requirement, we are free to choose any randomness that makes sense.

The most common random design aims to assemble a training set that contains samples that are uniformly distributed throughout the concentration space. Figure 5, shows such a training set. As compared to a factorially structrued training set, this type of randomly designed set will tend to have more samples in the central regions of the concentration space that at the perhiphery. This will tend to yield calibrations that have slightly better accuracy in predicting unknowns in the central regions than calibrations made with a factorial set, although the differences are usually slight.

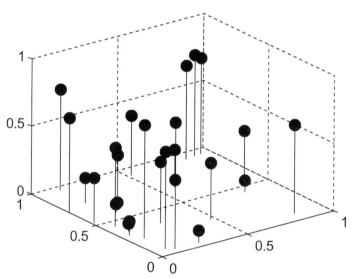

Figure 5. Randomly designed training set employing uniform distribution.

Another type of random design assembles samples that are normally distributed about one or more points in the concentration space. Such a training set is shown in Figure 6. The point that is chosen as the center of the normally distributed samples might, for example, be the location in the concentration where the operating point of a process is located. This would give us a training set with a population density that is greatest at the process operation point and declines in a gaussian fashion as we move away from the operating point. Since all of the chemometric techniques calculate calibrations that minimize the least squares errors at the calibration points, if we have a greater density of calibration samples in a particular region of the calibration space, the errors in this region will tend to be minimized at the expense of greater errors in the less densly populated regions. In this case, we would expect to get a calibration that would have maximum prediction accuracy for unknowns at the process operating point at the expense of the prediction accuracy for unknowns further away from the operating point.

Manually Designed Training Sets

There is nothing that says we must slavishly follow one of the structured or random experimental designs. For example, we might wish to combine the features of structured and random designs. Also, there are times when we have

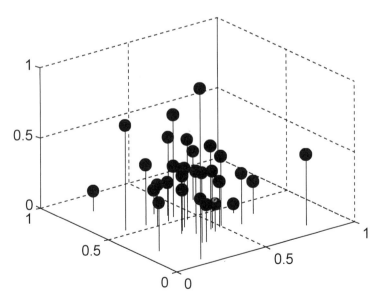

Figure 6. Randomly designed training set employing a normal distribution.

enough additional knowledge about an application that we can create a better
training set than any of the "canned" schemes would provide.

Manual design is most often used to augment a training set initially
constructed with the structured or random approach. Perhaps we wish to
enhance the accuray in one region of the calibration space. One way to do this is
to augment the training set with additional samples that occupy that region of
the space. Or perhaps we are concerned that a randomly designed training set
does not have adequate representation of samples at the perhiphery of the
calibration space. We could address that concern by augmenting the training set
with additional samples chosen by the factorial design approach. Figure 7
shows a training set that was manually augmented in this way. This give us the
advantages of both methods, and is a good way of including more samples in
the training set than is possible with a straight factorial design.

Finally, there are other times when circumstances do not permit us to freely
choose what we will use for calibration samples. If we are not able to dictate
what samples will go into our training set, we often must resort to the *TILI*
method. TILI stands for "take it or leave it." The TILI method must be
employed whenever the only calibration samples available are "samples of

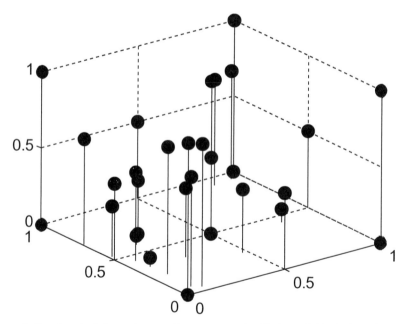

Figure 7. Random training set manually augmented with factorially designed samples.

opportunity." For example, we would be forced to use the TILI method whenever the only calibration samples available are the few specimens in the crumpled brown paper bag that the plant manager places on our desk as he explains why he needs a completely verified calibration within 3 days. Under such circumstances, success in never guaranteed. Any calibration created in this way would have to be used *very* carefully, indeed. Often, in these situations, the only responsible decision is to "leave it." It is better to produce no calibration at all rather than produce a calibration that is neither accurate nor reliable.

Creating the Training Set Concentration Matrices

We will now construct the concentration matrices for our training sets. Remember, we will simulate a 4-component system for which we have concentration values available for only 3 of the components. A random amount of the 4th component will be present in every sample, but when it comes time to generate the calibrations, we will not utilize any information about the concentration of the 4th component. Nonetheless, we must generate concentration values for the 4th component if we are to synthesize the spectra of the samples. We will simply ignore or discard the 4th component concentration values after we have created the spectra.

We will create 2 different training sets, one designed with the factorial structure including the mean concentration values, and one designed with a uniform random distribution of concentrations. We will not use the full-factorial structure. To keep our data sets smaller (and thus easier to plot graphically) we will eliminate those samples which lie on the midpoints of the edges of the calibration space. Each of the samples in the factorial training set will have a random amount of the 4th component determined by choosing numbers randomly from a uniform distribution of random numbers. Each of the samples in the random training set will have a random amount of each component determined by choosing numbers randomly from a uniform distribution of random numbers. The concentration ranges we use for each component are arbitrary. For simplicity, we will allow all of the concentrations to vary between a minimum of 0 and a maximum of 1 concentration unit.

We will organize the concentration values for the structured training set into a concentration matrix named **C1**. The concentrations for the randomly designed training set will be organized into a concentration matrix named **C2**. The factorial structured design for a 3-component system yields 15 different samples for **C1**. Accordingly, we will also assemble 15 different random samples in **C2**. Using column-wise data organization, **C1** and **C2** will each have 4 rows, one for each component, and 15 columns, one for each mixture. After we have constructed the absorbance spectra for the samples in **C1** and **C2**, we will discard the concentrations that are in the 4th row, leaving only the concentration values for the first 3 components. If you are already getting confused, remember that the table on the inside back cover summarizes all of the synthetic data we will be working with. Figure 8 contains multivariate plots of the concentrations of the 3 known components for each sample in **C1** and in **C2**.

Creating the Validation Set Concentration Matrices

Next, we create a concentration matrix containing mixtures that we will hold in reserve as validation data. We will assemble 10 different validation samples into a concentration matrix called **C3**. Each of the samples in this validation set will have a random amount of each component determined by choosing numbers randomly from a uniform distribution of random numbers between 0 and 1.

We will also create validation data containing samples for which the concentrations of the 3 known components are allowed to extend beyond the range of concentrations spanned in the training sets. We will assemble 8 of these overrange samples into a concentration matrix called **C4**. The concentration value for each of the 3 known components in each sample will be chosen randomly from a uniform distribution of random numbers between 0 and 2.5. The concentration value for the 4th component in each sample will be chosen randomly from a uniform distribution of random numbers between 0 and 1.

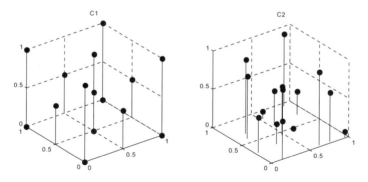

Figure 8. Concentration values for first 3 components of the 2 training sets.

We will create yet another set of validation data containing samples that have an additional component that was not present in any of the calibration samples. This will allow us to observe what happens when we try to use a calibration to predict the concentrations of an unknown that contains an unexpected interferent. We will assemble 8 of these samples into a concentration matrix called **C5**. The concentration value for each of the components in each sample will be chosen randomly from a uniform distribution of random numbers between 0 and 1. Figure 9 contains multivariate plots of the first three components of the validation sets.

Creating the Pure Component Spectra

We now have five different concentrations matrices. Before we can generate the absorbance matrices containing the spectra for all of these synthetic samples, we must first create spectra for each of the 5 pure components we are using: 3 components whose concentrations are known, a fourth component

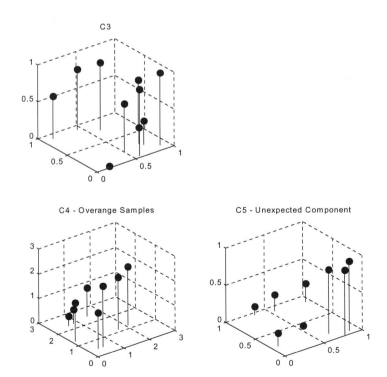

Figure 9. Concentration values for first 3 components of the validation sets.

which is present in unknown but varying concentrations, and a fifth component which is present as an unexpected interferent in samples in the validation set **C5**.

We will create the spectra for our pure components using gaussian peaks of various widths and intensities. We will work with spectra that are sampled at 100 discrete "wavelengths." In order to make our data realistically challenging, we will incorporate a significant amount of spectral overlap among the components. Figure 10 contains plots of spectra for the 5 pure components. We can see that there is a considerable overlap of the spectral peaks of Components 1 and 2. Similarly, the spectral peaks of Components 3 and 4 do not differ much in width or position. And Component 5, the unexpected interferent that is present in the 5th validation set, overlaps the spectra of all the other components. When we examine all 5 component spectra in a single plot, we can appreciate the degree of spectral overlap.

Creating the Absorbance Matrices — Matrix Multiplication

Now that we have spectra for each of the pure components, we can put the concentration values for each sample into the Beer-Lambert Law to calculate the absorbance spectrum for each sample. But first, let's review various ways of

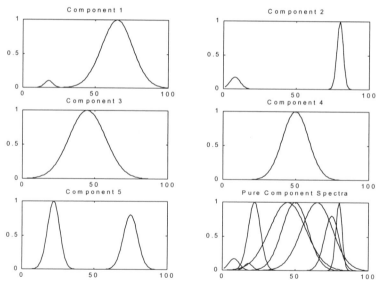

Figure 10. Synthetic spectra of the 5 pure components.

of representing the Beer-Lambert law. It is important that you are comfortable with the mechanics covered in the next few pages. In particular, you should make an effort to master the details of multiplying one matrix by another matrix. The mechanics of matrix multiplication are also discussed in Appendix A. You may also wish to consult other texts on elementary matrix algebra (see the bibliography) if you have difficulty with the approaches used here.

The absorbance at a single wavelength due to the presence of a single component is given by:

$$A = K\,C \qquad\qquad [19]$$

where:

A   is the absorbance at that wavelength

K   is the absorbance coefficient for that component and wavelength

C   is the concentration of the component

Please remember that even though we are using the vocabulary of spectroscopy, the concepts discussed here apply to any system where we can measure a quantity, A, that is proportional to some property, C, of our sample. For example, A could be the area of a chromatographic peak or the intensity of an elemental emission line, and C could be the concentration of a component in the sample.

Generalizing for multiple components and multiple wavelengths we get:

$$A_w = \sum_{c=1}^{n} K_{wc}\,C_c \qquad\qquad [20]$$

where:

A_w is the absorbance at the wth wavelength

K_{wc} is the absorbance coefficient at the wth wavelength for the cth component

C_c is the concentration of the cth component

n   is the total number or components

We can write equation [20] in expanded form:

$$
\begin{aligned}
A_1 &= K_{11}C_1 + K_{12}C_2 + \ldots + K_{1c}C_c \\
A_2 &= K_{21}C_1 + K_{22}C_2 + \ldots + K_{2c}C_c \\
A_2 &= K_{31}C_1 + K_{32}C_2 + \ldots + K_{3c}C_c \\
&\ldots \qquad \ldots \qquad \ldots \qquad \ldots \qquad \ldots \\
A_w &= K_{w1}C_1 + K_{w2}C_2 + \ldots + K_{wc}C_c
\end{aligned}
\qquad [21]
$$

We see from equation [21] that the absorbance at a given wavelength, w, is simply equal to the sum of the absorbances at that wavelength due to each of the components present.

We can also use the definition of matrix multiplication to write equation [21] as a matrix equation:

$$
\mathbf{A} = \mathbf{K}\,\mathbf{C} \qquad [22]
$$

where:

A is a single column absorbance matrix of the form of equation [1]

C is a single column concentration matrix of the form in equation [9]

K is a column-wise matrix of the form:

$$
\begin{bmatrix}
K_{11} & K_{12} & K_{13} & \ldots & K_{1s} \\
K_{21} & K_{22} & K_{23} & \ldots & K_{2s} \\
K_{31} & K_{32} & K_{33} & \ldots & K_{3s} \\
K_{41} & K_{42} & K_{43} & \ldots & K_{4s} \\
\ldots & \ldots & \ldots & \ldots & \ldots \\
K_{w1} & K_{w2} & K_{w3} & \ldots & K_{ws}
\end{bmatrix}
\qquad [23]
$$

If we examine the first column of the matrix in equation [23] we see that each K_{w1} is the absorbance at each wavelength, w, due to one concentration unit of component 1. Thus, the first column of the matrix is identical to the pure component spectrum of component 1. Similarly, the second column is identical to the pure component spectrum of component 2, and so on.

We have been considering equations [20] through [22] for the case where we are creating an absorbance matrix, **A**, that contains only a single spectrum organized as a single column vector in the matrix. **A** is generated by multiplying the pure component spectra in the matrix **K** by the concentration matrix, **C**, which contains the concentrations of each component in the sample. These concentrations are organized as a single column vector that corresponds to the single column vector in **A**. It is a simple matter to further generalize equation [20] to the case where we create an absorbance matrix, **A**, that contains any number of spectra, each held in a separate column vector in the matrix:

$$A_{ws} = \sum_{c=1}^{n} K_{wc} C_{cs} \qquad [24]$$

where:

A_{ws} is the absorbance at the wth wavelength for the sth sample

K_{wc} is the absorbance coefficient at the wth wavelength for the cth component and wavelength

C_{cs} is the concentration of the cth component for the sth sample

n   is the total number or components

In equation [24], **A** is generated by multiplying the pure component spectra in the matrix **K** by the concentration matrix, **C**, just as was done in equation [20]. But, in this case, **C** will have a column of concentration values for each sample. Each column of **C** will generate a corresponding column in **A** containing the spectrum for that sample. Note that equation [24] can also be written as equation [22]. We can represent equation [24] graphically:

$$[25]$$

Equation [25] shows an absorbance matrix containing the spectra of 4 mixtures. Each spectrum is measured at 15 different wavelengths. The matrix, **K**, is shown to hold the pure spectra of two different components, each measured at the 15 wavelengths. Accordingly, the concentration matrix must have 4 corresponding columns, one for each sample; and each column must have two concentration values, one for each component.

We can illustrate equation [25] in yet another way:

$$
\begin{array}{l}
\text{X X X X} \\
\text{X X X X} \\
\text{X X X X} \\
\text{X o X X} \\
\text{X X X X} \\
\text{X X X X} \\
\text{X X X X} \\
\text{X X X X} \\
\text{X X X X} \\
\text{X X X X} \\
\text{X X X X} \\
\text{X X X X} \\
\text{X X X X} \\
\text{X X X X} \\
\text{X X X X}
\end{array}
\quad = \quad
\begin{array}{l}
\text{X X} \\
\text{X X} \\
\text{X X} \\
\text{a\ \ b} \\
\text{X X} \\
\text{X X} \\
\text{X X} \\
\text{X X} \\
\text{X X} \\
\text{X X} \\
\text{X X} \\
\text{X X} \\
\text{X X} \\
\text{X X} \\
\text{X X}
\end{array}
\quad \times \quad
\begin{array}{l}
\text{X r X X} \\
\text{X s X X}
\end{array}
\qquad [26]
$$

We see in equation [26], for example, that the absorbance value in the 4th row and 2nd column of **A** is given by the vector multiplication of the 4th row of **K** with the 2nd column of **C**, thusly:

$$ o \; = \; (a \times r) \; + \; (b \times s) \qquad [27] $$

Again, please consult Appendix A if you are not yet comfortable with matrix multiplication.

Noise-Free Absorbance Matrices

So now we see that we can organize each of our 5 pure component spectra into a **K** matrix. In our case, the matrix will have 100 rows, one for each wavelength, and 5 columns, one for each pure spectrum. We can then generate an absorbance matrix for each concentration matrix, **C1** through **C5**, using equation [22]. We will name the resulting absorbance matrices **A1** through **A5**, respectively.

The spectra in each matrix are plotted in Figure 11. We can see that, at this point, the spectra are free of noise. Notice that the spectra in **A4**, which are the spectra of the overange samples, generally exhibit somewhat higher absorbances than the spectra in the other matrices. We can also see that the spectra in **A5**, which are the spectra of the samples with the unexpected 5th component, seem to contain some features that are absent from the spectra in the other matrices.

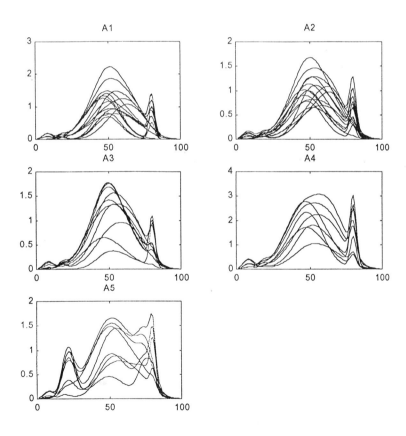

Adding Realism

Unfortunately, real data is never as nice as this perferctly linear, noise-free data that we have just created. What's more, we can't learn very much by experimenting with data like this. So, it is time to make this data more realistic. Simply adding noise will not be sufficient. We will also add some artifacts that are often found in data collected on real instruments from actual industrial samples.

Adding Baselines

All of the spectra are resting on a flat baseline equal to zero. Most real instruments suffer from some degree of baseline error. To simulate this, we will add a different random amount of a linear baseline to each spectrum. Each baseline will have an offset randomly chosen between .02 and -.02, and a slope randomly chosen between .2 and -.2. Note that these baselines are not completely realistic because they are perfectly straight. Real instruments will often produce baselines with some degree of curvature. It is important to understand that baseline curvature will have the same effect on our data as would the addition of varying levels of an unexpected interfering component that was not included in the training set. We will see that, while the various calibration techniques are able to handle perfectly straight baselines rather well, to the extent an instrument introduces a significant amount of nonreproducible baseline curvature, it can become difficult, if not impossible, to develop a useable calibration for that instrument. The spectra with added linear baselines are plotted in Figure 12.

Adding Non-Linearities

Nearly all instrumental data contain some nonlinearities. It is only a question of how much nonlinearity is present. In order to make our data as realistic as possible, we now add some nonlinearity to it. There are two major sources of nonlinearities in chemical data:

1. Instrumental

2. Chemical and physical

Chemical and physical nonlinearities are caused by interactions among the components of a system. They include such effects as peak shifting and broadening as a function of the concentration of one or more components in the sample. Instrumental nonlinearities are caused by imperfections and/or nonideal behavior in the instrument. For example, some detectors show a

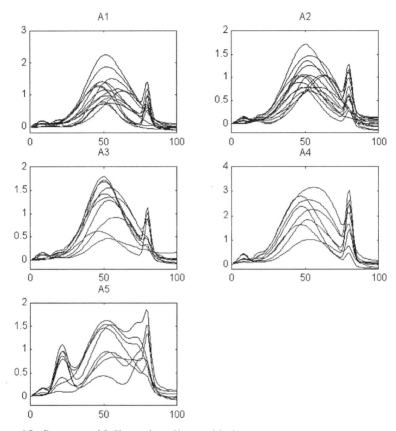

Figure 12. Spectra with linear baselines added.

saturation effect that reduces the response to a signal as the signal level increases. Figure 13 shows the difference in response between a perfectly linear detector and one with a 5% quadratic nonlinearity.

We will add a 1% nonlinear effect to our data by reducing every absorbance value as follows:

$$A_{nonlinear} = A - .01\ A^2 \qquad [28]$$

Where:

$A_{nonlinear}$ is the new value of the absorbance with the nonlinearity

A is the original value of the absorbance

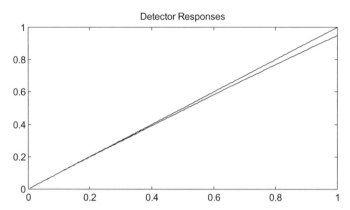

Figure 13. Response of a linear (upper) and a 5% nonlinear (lower) detector.

1% is a significant amount of nonlinearity. It will be interesting to observe the impact the nonlinearity has on our calibrations. Figure 14 contains plots of **A1** through **A5** after adding the nonlinearity. There aren't any obvious differences between the spectra in Figure 12 and Figure 14. The last panel in Figure 14 shows a magnified region of a single spectrum from **A1** plotted before and after the nonlinearity was incorporated into the data. When we plot at this magnification, we can now see how the nonlinearity reduces the measured response of the absorbance peaks.

Adding Noise

The last elements of realism we will add to the data is random error or noise. In actual data there is noise both in the measurement of the spectra, and in the determination of the concentrations. Accordingly, we will add random error to the data in the absorbance matrices and the concentration matrices.

Concentration Noise

We will now add random noise to each concentration value in **C1** through **C5**. The noise will follow a gaussian distribution with a mean of 0 and a standard deviation of .02 concentration units. This represents an average relative noise level of approximately 5% of the mean concentration values — a level typically encountered when working with industrial samples. Figure 15 contains multivariate plots of the noise-free and the noisy concentration values for **C1** through **C5**. We will not make any use of the noise-free concentrations since we never have these when working with actual data.

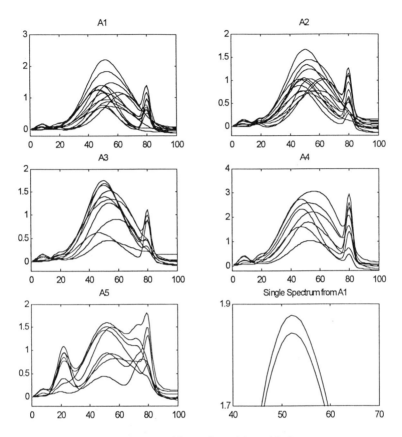

Figure 14. Absorbance spectra with nonlinearities added.

Absorbance Noise

In a similar fashion, we will now add random noise to each absorbance value in **A1** through **A5**. The noise will follow a gaussian distribution with a mean of 0 and a standard deviation of .05 absorbance units. This represents a relative noise level of approximately 10% of the mean absorbance values. This noise level is high enough to make the calibration realistically challenging — a level typically encountered when working wth industrial samples. Figure 16 contains plots of the resulting spectra in **A1** through **A5**. We can see that the noise is high enough to obscure the lower intensity peaks of components 1 and 2. We will be working with these noisy spectra throughout the rest of this book.

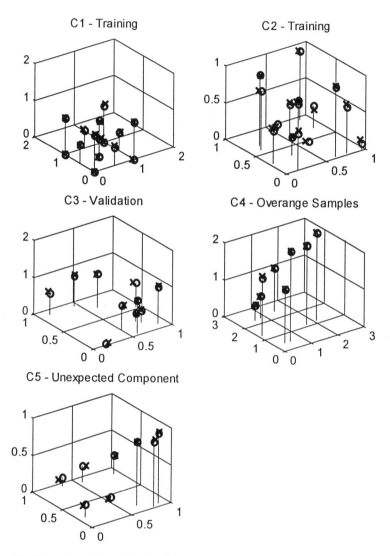

Figure 15. Noise-free (O) and noisy (x) concentration values.

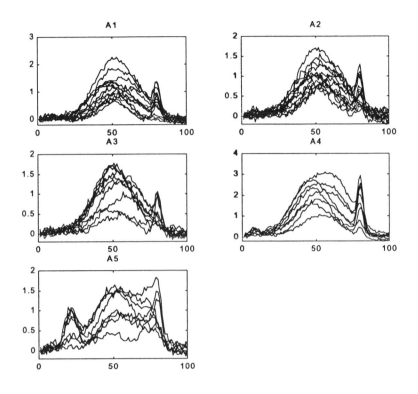

Figure 16. Absorbance spectra with noise added.

Classical Least-Squares

Classical least-squares (CLS), sometimes known as K-matrix calibration, is so called because, originally, it involved the application of multiple linear regression (MLR) to the classical expression of the Beer-Lambert Law of spectroscopy:

$$A = K\,C \qquad\qquad [1]$$

This is the same equation we used to create our simulated data. We discussed it thoroughly in the last chapter. If you have "just tuned in" at this point in the story, you may wish to review the discussion of equations [19] through [27] before continuing here.

Computing the Calibration

To produce a calibration using classical least-squares, we start with a training set consisting of a concentration matrix, C, and an absorbance matrix, A, for known calibration samples. We then solve for the matrix, K. Each column of K will each hold the spectrum of one of the pure components. Since the data in C and A contain noise, there will, in general, be no exact solution for equation [29]. So, we must find the best least-squares solution for equation [29]. In other words, we want to find K such that the sum of the squares of the errors is minimized. The errors are the difference between the measured spectra, A, and the spectra calculated by multiplying K and C:

$$\text{errors} = K\,C \; - \; A \qquad\qquad [2]$$

To solve for K, we first post-multiply each side of the equation by C^T, the transpose of the concentration matrix.

$$A\,C^T = K\,C\,C^T \qquad\qquad [3]$$

Recall that the matrix C^T is formed by taking every row of C and placing it as a column in C^T. Next, we eliminate the quantity $[C\,C^T]$ from the right-hand side of equation [31]. We can do this by post-multiplying each side of the equation by $[C\,C^T]^{-1}$, the matrix inverse of $[C\,C^T]$.

$$A \; C^T \; [C \; C^T]^{-1} = K \; [C \; C^T] \; [C \; C^T]^{-1} \qquad [32]$$

$[C \; C^T]^{-1}$ is known as the pseudo inverse of C. Since the product of a matrix and its inverse is the identity matrix, $[C \; C^T][C \; C^T]^{-1}$ disappears from the right-hand side of equation [32] leaving

$$A \; C^T \; [C \; C^T]^{-1} = K \qquad [33]$$

In order for the inverse of $[C \; C^T]$ to exist, C must have at least as many columns as rows. Since C has one row for each component and one column for each sample, this means that we must have at least as many samples as components in order to be able to compute equation [33]. This would certainly seem to be a reasonable constraint. Also, if there is any linear dependence among the rows or columns of C, $[C \; C^T]$ will be singular and its inverse will not exist. One of the most common ways of introducing linear dependency is to construct a sample set by serial dilution.

Predicting Unknowns

Now that we have calculated K we can use it to predict the concentrations in an unknown sample from its measured spectrum. First, we place the spectrum into a new absorbance matrix, A_{unk}. We can now use equation [29] to give us a new concentration matrix, C_{unk}, containing the predicted concentration values for the unknown sample.

$$A_{unk} = K \; C_{unk} \qquad [34]$$

To solve for C_{unk}, we first pre-multiply both sides of the equation by K^T.

$$K^T \, A_{unk} = K^T \; K \; C_{unk} \qquad [35]$$

Next, we eliminate the quantity $[K^T \; K]$ from the right-hand side of equation [35]. We can do this by pre-multiplying each side of the equation by $[K^T \; K]^{-1}$, the matrix inverse of $[K^T \; K]$.

$$[K^T \; K]^{-1} \; K^T \, A_{unk} = [K^T \; K]^{-1} \; [K^T \; K] \; C_{unk} \qquad [36]$$

$[\mathbf{K^T\,K}]^{-1}$ is known as the pseudo inverse of $\mathbf{K}$. Since the product of a matrix and its transpose is the identity matrix, $[\mathbf{K^T\,K}]^{-1}[\mathbf{K^T\,K}]$ disappears from the right-hand side of equation [36] leaving

$$[\mathbf{K^T\,K}]^{-1}\,\mathbf{K^T\,A_{unk}} = \mathbf{C_{unk}} \qquad [37]$$

In order for the inverse of $[\mathbf{K^T\,K}]$ to exist, $\mathbf{K}$ must have at least as many rows as columns. Since $\mathbf{K}$ has one row for each wavelength and one column for each component, this means that we must have at least as many wavelengths as components in order to be able to compute equation [37]. This constraint also seems reasonable.

Taking advantage of the associative property of matrix multiplication, we can compute the quantity $[\mathbf{K^T\,K}]^{-1}\,\mathbf{K^T}$ at calibration time.

$$\mathbf{K_{cal}} = [\mathbf{K^T\,K}]^{-1}\,\mathbf{K^T} \qquad [38]$$

$\mathbf{K_{cal}}$ is called the calibration matrix or the regression matrix. It contains the calibration, or regression, coefficients which are used to predict the concentrations of an unknown from its spectrum. $\mathbf{K_{cal}}$ will contain one row of coefficients for each component being predicted. Each row will have one coefficient for each spectral wavelength. Thus, $\mathbf{K_{cal}}$ will have as many columns as there are spectral wavelengths. Substituting equation [38] into equation [37] gives us

$$\mathbf{C_{unk}} = \mathbf{K_{cal}\,A_{unk}} \qquad [39]$$

Thus, we can predict the concentrations in an unknown by a simple matrix multiplication of a calibration matrix and the unknown spectrum.

Additional Constraints

We have already noted that CLS requires at least as many samples and at least as many wavelengths as there are components. These constraints seem perfectly reasonable. But, when we use CLS, we must also satisfy another requirement that gives cause for concern.

This requirement becomes apparent when we examine equation [21], which is reproduced, below, as equation [40].

$$
\begin{aligned}
A_1 &= K_{11}C_1 + K_{12}C_2 + \ldots + K_{1c}C_c \\
A_2 &= K_{21}C_1 + K_{22}C_2 + \ldots + K_{2c}C_c \\
A_3 &= K_{31}C_1 + K_{32}C_2 + \ldots + K_{3c}C_c \qquad [22] \\
& \ldots \qquad \ldots \qquad \ldots \qquad \ldots \\
A_w &= K_{w1}C_1 + K_{w2}C_2 + \ldots + K_{wc}C_c
\end{aligned}
$$

Equation [40] asserts that we are fully reconstructing the absorbance, A, at each wavelength. In other words, we are stating that we will account for all of the absorbance at each wavelength in terms of the concentrations of the components present in the sample. This means that, when we use CLS, we assume that we can provide accurate concentration values for *all* of the components in the sample. We can easily see that, when we solve for K for any component in equation [40], we will get an expression that includes the concentrations of all of the components.

It is usually difficult, if not impossible, to quantify all of the components in our samples. This is expecially true when we consider the meaning of the word "components" in the broadest sense. Even if we have accurate values for all of the constituents in our samples, how do we quantify the contribution to the spectral absorbance due to instrument drift, operator effect, instrument aging, sample cell alignment, etc.? The simple answer is that, generally, we can't. To the extent that we do not provide CLS with the concentration of all of the components in our samples, we might expect CLS to have problems. In the case of our simulated data, we have samples that contain 4 components, but we only have concentration values for 3 of the components. Each sample also contains a random baseline for which "concentration" values are not available. Let's see how CLS handles these data.

CLS Results

We now use CLS to generate calibrations from our two training sets, **A1** and **A2**. For each training set, we will get matrices, **K1** and **K2**, respectively, containing the best least-squares estimates for the spectra of pure components 1 - 3, and matrices, **K1$_{cal}$** and **K2$_{cal}$**, each containing 3 rows of calibration coefficients, one row for each of the 3 components we will predict. First, we will compare the estimated pure component spectra to the actual spectra we started with. Next, we will see how well each calibration matrix is able to predict the concentrations of the samples that were used to generate that calibration. Finally, we will see how well each calibration is able to predict the concentrations of the unknown samples contained in the three validation sets, **A3** through **A5**.

As we've already noted, the most difficult part of this work is keeping track of which data and which results are which. If you find yourself getting confused, you may wish to consult the data "crib sheet" located on the inside back cover.

Estimated Pure Component Spectra

Figure 17 contains plots of the pure component spectra calculated by CLS together with the actual pure component spectra we started with. The smooth curves are the actual spectra, and the noisy curves are the CLS estimates. Since we supplied concentration values for 3 components, CLS returns 3 estimated pure component spectra. The left-hand column of Figure 17 contains the spectra calculated from **A1**, the training set with the structured design. The right-hand column of Figure 17 contains the spectra calculate from **A2**, the training set with the random design.

We can see that the estimated spectra, while they come close to the actual spectra, have some significant problems. We can understand the source of the problems when we look at the spectrum of Component 4. Because we stated in equation [40] that we will account for all of the absorbance in the spectra, CLS was forced to distribute the absorbance contributions from Component 4 among the other components. Since there is no "correct" way to distribute the Component 4 absorbance, the actual distribution will depend upon the makeup of the training set. Accordingly, we see that CLS distributed the Component 4 absorbance differently for each training set. We can verify this by taking the sum of the 3 estimated pure component spectra, and subtracting from it the sum of the actual spectra of the first 3 components:

$$K_{residual} = (K_1 + K_2 + K_3) - (A1_{pure} + A2_{pure} + A3_{pure}) \qquad [41]$$

where:

K_1, K_2, K_3	are the estimated pure component spectra (the columns of **K)** for Components 1 - 3, respectively;
$A1_{pure}$, $A2_{pure}$, $A3_{pure}$	are the actual spectra for Components 1 - 3.

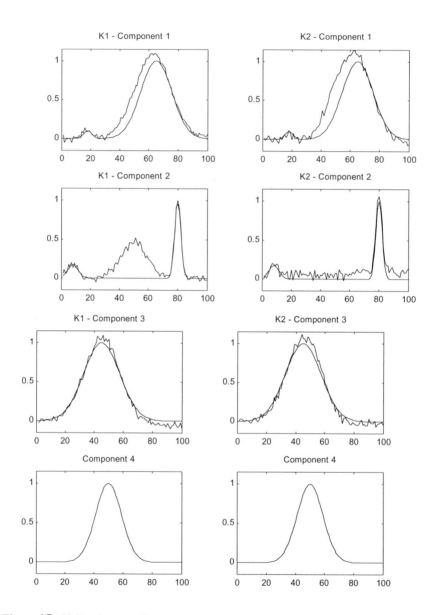

Figure 17. CLS estimates of pure component spectra.

These $K_{residual}$ (noisy curves) for each training set are plotted in Figure 18 together with the actual spectrum of Component 4 (smooth curves).

Returning to Figure 17, it is interesting to note how well CLS was able to estimate the low intensity peaks of Components 1 and 2. These peaks lie in an area of the spectrum where Component 4 does not cause interference. Thus, therewas no distribution of excess absorbance from Component 4 to disrupt the estimate in that region of the spectrum. If we look closely, we will also notice that the absorbance due to the sloping baselines that we added to the simulated data has also been distributed among the estimated pure component spectra. It is particularly visible in **K1**, Component 3 and **K2** Component 2.

Fit to the Training Set

Next, we examine how well CLS was able to fit the training set data. To do this, we use the CLS calibration matrix K_{cal} to predict (or estimate) the concentrations of the samples with which the calibration was generated. We then examine the differences between these predicted (or estimated) concentrations and the actual concentrations. Notice that "predict" and "estimate" may be used interchangeably in this context. We first substitute $K1_{cal}$ and **A1** into equation [39], naming the resulting matrix with the predicted concentrations $K1_{res}$. We then repeat the process with $K2_{cal}$ and A2, naming the resulting concentration matrix $K2_{res}$.

Figure 19 contains plots of the expected (x-axis) vs. predicted (y-axis) concentrations for the fits to training sets **A1** and **A2**. (Notice that the expected concentration values for **A1**, the factorially designed training set are either 0.0, 0.5, or 1.0, plus or minus the added noise). While there is certainly a recognizable correlation between the expected and predicted concentration values this is not as good a fit as we might have hoped for.

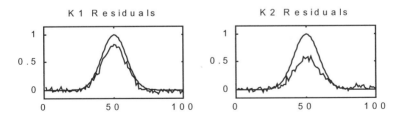

Figure 18. Residuals of estimated pure component spectra (see text).

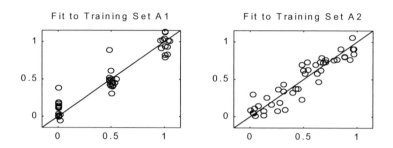

Figure 19. Expected concentrations (x-axis) vs. predicted concentrations (y-axis) for the fit to training sets **A1** and **A2**.

It is very important to understand that these fits only give us an indication of how well we are able to fit the calibration data with a linear regression. A good fit to the training set does not guarantee that we have a calibration with good predictive ability. All we can conclude, in general, from the fits is that we would expect that a calibration would not be able to predict the concentrations of unknowns more precisely than it is able to fit the training samples. If the fit to the training data is generally poor, as it is here, it could be caused by large errors in the expected concentration values as determined by the referee method. We know that this can't be the case for our data. The problem, in this case, is most due to the presence of varying amounts of the fourth component for which concentration values are unavailable.

Predictions on Validation Set

To draw conclusions about how well the calibrations will perform on unknown samples, we must examine how well they can predict the concentrations in our 3 validation sets **A3 - A5**. We do this by substituting **A3 - A5** into equation [39], first with $K1_{cal}$, then with $K2_{cal}$ to produce 6 concentration matrices containing the estimated concentrations. We will name these matrices $K13_{res}$ through $K15_{res}$ and $K23_{res}$ through $K25_{res}$. Using this naming system, $K24_{res}$ is a concentration matrix holding the concentrations for validation set **A4** predicted with the calibration matrix $K2_{cal}$, that was generated with training set **A2**, the one which was constructed with the random design. Figure 20 contains plots of the expected vs. predicted concentrations for $K13_{res}$ through $K25_{res}$.

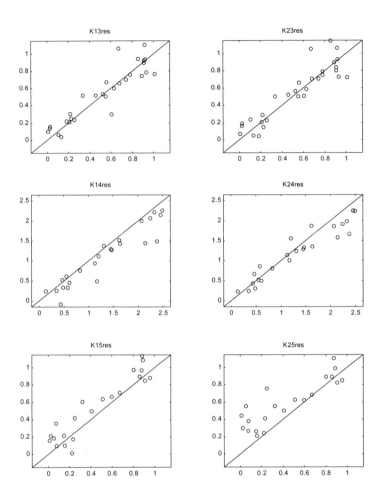

Figure 20. Expected concentrations (x-axis) vs. predicted concentrations (y-axis) for $K13_{res}$ through $K23_{res}$ (see text).

We can see, in Figure 20 that we get similar results when we use the two calibrations, $K1_{cal}$ and $K2_{cal}$, to predict the concentrations in the validation sets. When we examine the plots for $K13_{res}$ and $K23_{res}$, the predictions for our normal validation set, $A3$, we see that, while the calibrations do work to a certain degree, there is a considerable amount of scatter between the expected and the predicted values. For some applications, this might be an acceptable level of performance. But, in general, we would hope to do much better.

$K14_{res}$ and $K24_{res}$, the predictions for the validation set, **A4**, whose samples contain some overrange concentration values show a similar degree of scatter. But remember that the scale of these two plots is larger and the actual magnitude of the errors is correspondingly larger. We can also see a curvature in the plots. The predicted values at the higher concentration levels begin to drop below the ideal regression line. This is due to the nonlinearity in the absorbance values which diminishes the response of the higher concentration samples below what they would otherwise be if there were no nonlinearity.

$K15_{res}$ and $K25_{res}$, the predictions for the validation set, **A5**, whose samples contain varying amounts of a 5th component that was never present in the training sets, are surprisingly good when compared to $K13_{res}$ and $K23_{res}$. But this is more an indication of how bad $K13_{res}$ and $K23_{res}$ are rather than how good $K15_{res}$ and $K25_{res}$ are. In any case, these results are not to be trusted. Whenever a new interfering component turns up in an unknown sample, the calibration must be considered invalid. Unfortunatley, neither CLS nor ILS can provide any direct indication that this condition might exist.

We can also examine these results numerically. One of the best ways to do this is by examining the Predicted Residual Error Sum-of-Squares or PRESS. To calculate PRESS we compute the errors between the expected and predicted values for all of the samples, square them, and sum them together.

$$\text{PRESS} \ = \ \Sigma \ (C_{predicted} - C_{expected})^2 \qquad\qquad [42]$$

Usually, PRESS should be calculated separately for each predicted component, and the calibration optimized individually for each component. For preliminary work, it can be convenient to calculate PRESS collectively for all components together, although it isn't always possible to do so if the units for each component are drastically different or scaled in drastically different ways. Calculating PRESS collectively will be sufficient for our purposes. This will give us a single PRESS value for each set of results $K1_{res}$ through $K25_{res}$. Since not all of the data sets have the same number of samples, we will divide each of these PRESS values by the number of samples in the respective data sets so that they can be more directly compared. We will also divide each value by the number of components predicted (in this case 3). The resulting press values are compiled in Table 2.

Strictly speaking, this is not a correct way to normalize the PRESS values when not all of the data sets contain the same number of samples. If we want to

	$K1_{cal}$			$K2_{cal}$	
	PRESS	SEC²	r	PRESS	SEC²
A1	.0191	.0204	.9456	-	-	-
A2	-	-	-	.0127	.0159	.9310
A3	.0171	.0143	.9091	.0188	.0173	.9100
A4	.0984	.0745	.9696	.0697	.0708	.9494
A5	.0280	.0297	.9667	.0744	.0688	.9107

Table 2. PRESS, SEC², SEP², and r for $K1_{res}$ through $K25_{res}$.

correctly compare PRESS values for data sets that contain differing numbers of samples, we should convert them to Standard Error of Calibration (SEC), sometimes called the Standard Error of Estimate (SEE), for the training sets, and Standard Error of Prediction (SEP) for the validation sets. A detailed discussion of SEC, SEE and SEP can be found in Appendix B. As we can see in Table 2, in this case, dividing PRESS by the number of samples and components give us a value that is almost the same as the SEC and SEP values.

It is important to realize that there are often differences in the way the terms PRESS, SEC, SEP, and SEE are used in the literature. Errors in usage also appear. Whenever you encounter these terms, it is necessary to read the article carefully in order to understand exactly what they mean in each particular publication. These terms are discussed in more detail in Appendix II.

Table 2 also contains the correlation coefficient, r, for each K_{res}. If the predicted concentrations for a data set exactly matched the expected concentrations, r would equal 1.0. If there were absolutely no relationship between the predicted and expected concentrations, r would equal 0.0.

The Regression Coefficients

It is also interesting to examine the actual regression coefficients that each calibration produces. Recall that we get one row in the calibration matrix, K_{cal}, for each component that is predicted. Each row contains one coefficient for each wavelength. Thus, we can conveniently plot each row of K_{cal} as if it were a spectrum. Figure 21 contains a set of such plots for each component for $K1_{cal}$ and $K2_{cal}$. We can think of these as plots of the "strategy" of the calibration showing which wavelengths are used in positive correlation, and which in negative correlation.

We see, in Figure 21 , that the strategy for component 1 is basically the same for the two calibrations. But, there are some striking differences between the two calibrations for components 2 and 3. A theoretical statistician might suggest that each of the different strategies for the different components is equally statistically valid, and that, in general, there is not necessarily *a single* best calibration but may be, instead, a *plurality* of possible calibrations whose performances, one from another, are statistically indistinguishable. But, an analytical practitioner would tend to be uncomfortable whenever changes in the makeup of the calibration set cause significant changes in the resulting calibrations.

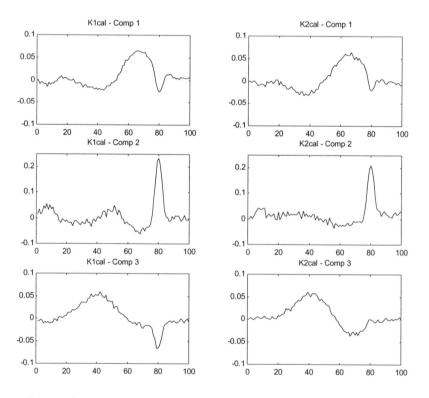

Figure 21. Plots of the CLS calibration coefficients calculated for each component with each training set.

CLS with Non-Zero Intercepts

There are any number of variations that can be applied to the CLS technique. Here we will only consider the most important one: non-zero intercepts. If you are interested in some of the other variations, you may wish to consult the references in the CLS section of the bibliography.

Referring to equation [40], we can see that we require the absorbance at each wavelength to equal zero whenever the concentrations of all the components in a sample are equal to zero. We can add some flexibility to the CLS calibration by eliminating this constraint. This will add one additional degree of freedom to the equations. To allow these non-zero intercepts, we simply rewrite equation [40] with a constant term for each wavelength:

$$
\begin{aligned}
A_1 &= K_{11}C_1 + K_{12}C_2 + \ldots + K_{1c}C_c + G_1 \\
A_2 &= K_{21}C_1 + K_{22}C_2 + \ldots + K_{2c}C_c + G_2 \\
A_3 &= K_{31}C_1 + K_{32}C_2 + \ldots + K_{3c}C_c + G_3 \qquad [43] \\
&\quad \ldots \qquad \ldots \qquad \ldots \qquad \ldots \qquad \ldots \\
A_w &= K_{w1}C_1 + K_{w2}C_2 + \ldots + K_{wc}C_c + G_w
\end{aligned}
$$

We have named the constant term G to emphasize that adding a constant term provides CLS a place to discard the "garbage," i.e. that portion of the absorbance at each wavelength that doesn't correlate well with the concentrations of the various components. Equation [43] still requires that we account for all of the absorbances in the training set spectra. But, now we are no longer required to distribute "spurious" absorbances from baseline effects, additional components, etc., among the estimated pure component spectra of the components we are trying to predict. Rewriting equation [43] in slightly greater detail:

$$
\begin{aligned}
A_1 &= K_{11}C_1 + K_{12}C_2 + \ldots + K_{1c}C_c + G_1C_g \\
A_2 &= K_{21}C_1 + K_{22}C_2 + \ldots + K_{2c}C_c + G_2C_g \\
A_3 &= K_{31}C_1 + K_{32}C_2 + \ldots + K_{3c}C_c + G_3C_g \qquad [44] \\
&\quad \ldots \qquad \ldots \qquad \ldots \qquad \ldots \qquad \ldots \\
A_w &= K_{w1}C_1 + K_{w2}C_2 + \ldots + K_{wc}C_c + G_wC_g
\end{aligned}
$$

we see that each constant term G_w, is actually being multiplied by some concentration term C_g which is completely arbitrary, although it must be constant for all of the samples in the training set. It is convenient to set C_g to unity. Thus, we have added an additional "component" to our training sets whose concentration is always equal to unity. So, to calculate a CLS calibration with nonzero intercepts, all we need to do is add a row of 1^{rs} to our original training set concentration matrix.

$$
\begin{array}{cccc}
C_{11} & C_{12} & \cdots & C_{1s} \\
C_{21} & C_{22} & \cdots & C_{2s} \\
\cdots & \cdots & \cdots & \cdots \\
C_{c1} & C_{c2} & \cdots & C_{cs} \\
1 & 1 & 1 & 1
\end{array}
\qquad [45]
$$

This will cause CLS to calculate an additional pure component spectrum for the G^{rs}. It will also give us an additional row of regression coefficients in our calibration matrix, $\mathbf{K_{cal}}$, which we can, likewise, discard.

Let's examine the results we get from a CLS calibration with nonzero intercept. We will use the same naming system we used for the first set of CLS results, but we will append an "a" to every name to designate the case of non-zero intercept. Thus, the calibration matrix calculated from the first training set will be named $\mathbf{K1a_{cal}}$, and the concentrations predicted for $\mathbf{A4}$, the validation set with the overrange concentration values will be held in a matrix named $\mathbf{K14a_{res}}$. If you aren't yet confused by all of these names, just wait, we've only begun. Figure 22 contains plots of the estimated pure component spectra for the 2 calibrations. We also plot the "pure spectrum" estimated by each calibration for the Garbage variable. Recall that each pure component spectrum is a column in the K matrices $\mathbf{K1a}$ and $\mathbf{K2a}$.

Examining Figure 22, we see that Garbage spectrum has, indeed, provided a place for CLS to discard extraneous absorbances. Note the similarity between the Garbage spectra in Figure 22 and the residual spectra in Figure 18. We can also see that CLS now does rather well in estimating the spectrum of Component 1. The results for Component 2 are a bit more mixed. The calibration on the first training set yields a better spectrum this time, but the calibration on the second training set yields a spectrum that is about the same, or perhaps a bit worse. And the spectra we get for Component 3 from both training sets do not appear to be as good as the spectra from the original zero-intercept calibration.

But the nonzero intercepts also allow an additional degree of freedom when we calculate the calibration matrix, $\mathbf{K_{cal}}$. This provides additional opportunity to adjust to the effects of the extraneous absorbances.

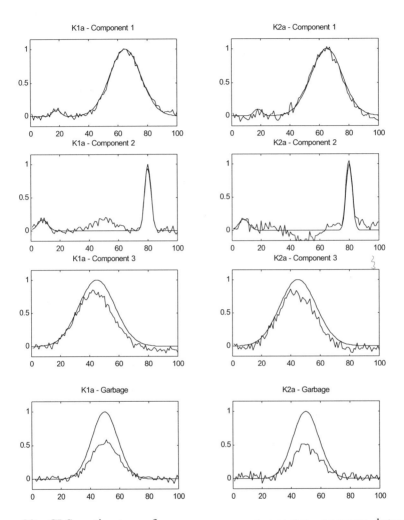

Figure 22. CLS estimates of pure component spectra, nonzero intercept calibration.

Figure 23 contains plots of the expected vs. predicted concentrations for all of the nonzero intercept CLS results. We can easily see that these results are much better than the results of the first calibrations. It is also apparent that when we predict the concentrations from the spectra in A5, the validation set with the

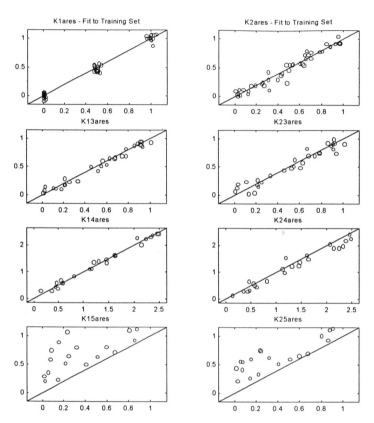

Figure 23. Expected concentrations (x-axis) vs. predicted concentrations (y-axis) for nonzero intercept CLS calibrations (see text).

unexpected 5th component, the results are, as expected, nearly useless. We can now appreciate the value of allowing nonzero intercepts when doing CLS. Especially so when we recall that, even if we know the concentrations of all the constituents in our samples, we are not likely to have good "concentration" values for baseline drift and other sources of extraneous absorbance in our spectra.

To complete the story, Table 3 contains the values for PRESS, SEC^2, SEP^2, and r, for this set of results.

	$K1a_{cal}$			$K2a_{cal}$		
	PRESS	SEC²	r	PRESS	SEC²	r
A1	.0026	.0034	.9924	-	-	-
A2	-	-	-	.0052	.0059	.9723
A3	.0030	.0033	.9844	.0074	.0075	.9622
A4	.0084	.0089	.9934	.0294	.0297	.9781
A5	.1763	.1920	.8576	.1148	.1261	.9016

Table 3. PRESS, SEC², SEP², and r for $K1a_{res}$ through $K25a_{res}$.

Some Easier Data

It would be interesting to see how well CLS would have done if we hadn't had a component whose concentration values were unknown (Component 4). To explore this, we will create two more data sets, **A6**, and **A7**, which will not contain Component 4. Other than the elimination of the 4th component, **A6** will be identical to **A2**, the randomly structured training set, and **A7** will be identical to **A3**, the normal validation set. The noise levels in **A6**, **A7**, and their corresponding concentration matrices, **C6** and **C7**, will be the same as in **A2**, **A3**, **C2**, and **C3**. But, the actual noise will be newly created—it won't be the exact same noise. The amount of nonlinearity will be the same, but since we will not have any absorbances from the 4th component, the impact of the nonlinearity will be slightly less. Figure 24 contains plots of the spectra in **A6** and **A7**.

We perform CLS on A6 to produce 2 calibrations. **K6** and $K6_{cal}$ are the matrices holding the pure component spectra and calibration coefficients, respectively, for CLS with zero intercepts. **K6a** and $K6a_{cal}$ are the corresponding matrices for CLS with nonzero intercepts.

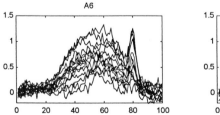
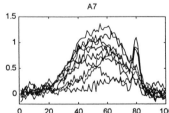

Figure 24. Absorbance spectra with noise added.

Figure 25 contains plots of the pure component spectra for the two calibrations. It is apparent that, in the absence of the extraneous absorbances from Component 4, CLS is not able to do a good job of estimating the pure component spectra. However, even with nonzero intercepts, CLS is unable to remove the sloping baseline from the spectra. Both calibrations distributed most of the baseline effect onto the spectrum for Component 2 and some onto the Component 3 spectrum.

Figure 26 contains plots of the expected vs. predicted concentrations using $K6_{cal}$ and $K6a_{cal}$. We see that the results are now much better. Notice that the predictions, $K63_{res}$ and $K63a_{res}$, for the original validation set A3, show a lot of points with large errors. This is exactly what we would expect when we recall that A3 has spectral contributions from Component 4. Thus, the samples in A3 contain varying amounts of an unexpected interferent that was not present in the training set.

Table 4 contains the values for PRESS, SEC^2, SEP^2, and r, for this set of results. We see from these last results that CLS can work quite well under the right conditions. In particular, it is important that we provide concentration values for all of the components present in the training sample.

Recognizing the difficulty satisfying the requirements for successful CLS, you may wonder why anyone would ever use CLS. There are a number of applications where CLS is particularly appropriate. One of the best examples is the case where a library of quantitative spectra is available, and the application requires the analysis of one or more components that suffer little or no interference other than that caused by the components themselves. In such cases, we do not need to use equation [33] to calculate the pure component spectra if we already have them in a library. We can simply construct a **K** matrix containing the required library spectra and proceed directly to equation [34] to calculate the calibration matrix K_{cal}.

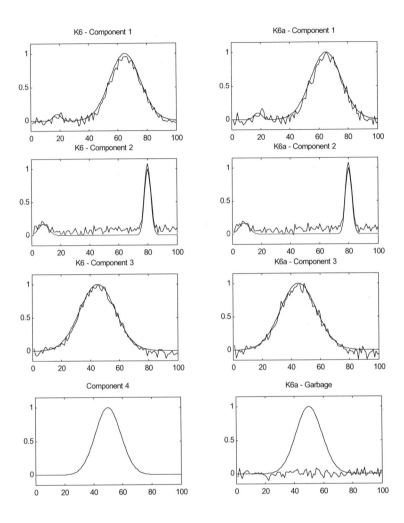

Figure 25. CLS estimates of pure component spectra.

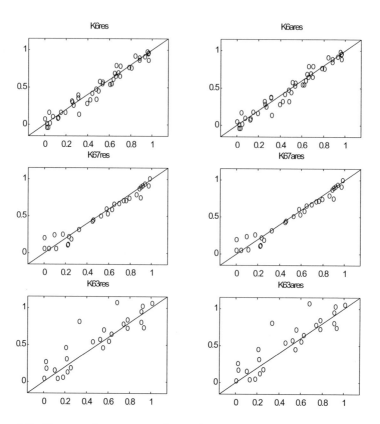

Figure 26. Expected concentrations (x-axis) vs. predicted concentrations (y-axis) for CLS calibrations (see text).

	$K6_{cal}$			$K6a_{cal}$		
A6	.0039	.0034	.9924	.0039	.0059	.9723
A7	.0048	.0033	.9844	.0046	.0075	.9622
A3	.0599	.0089	.9934	.0591	.0297	.9781

Table 4. PRESS, SEC^2, SEP^2, and r for $K6_{cal}$ and $K6a_{cal}$ results.

Inverse Least-Squares

Inverse least-squares (CLS), sometimes known as P-matrix calibration, is so called because, originally, it involved the application of multiple linear regression (MLR) to the inverse expression of the Beer-Lambert Law of spectroscopy:

$$C = P A \qquad [46]$$

Equation [46] is a matrix equation. Notice the similarity of this equation and equation [39]. For clarity, we can expand this equation to give:

$$
\begin{aligned}
C_1 &= P_{11}A_1 + P_{12}A_2 + \ldots + P_{1w}A_w \\
C_2 &= P_{21}A_1 + P_{22}A_2 + \ldots + P_{2w}A_w \\
C_3 &= P_{31}A_1 + P_{32}A_2 + \ldots + P_{3w}A_w \\
&\ldots \quad\quad \ldots \quad\quad \ldots \quad\quad \ldots \quad\quad \ldots \\
C_c &= P_{c1}A_1 + P_{w2}A_2 + \ldots + P_{cw}A_w
\end{aligned}
\qquad [47]
$$

where:

A_w is the absorbance at the wth wavelength

P_{cw} is the calibration coefficient for the cth component at the wth wavelength

C_c is the concentration of the cth component

Computing the Calibration

To produce a calibration using inverse least-squares, we start with a training set consisting of a concentration matrix, **C**, and an absorbance matrix, **A**, for known calibration samples. We then solve for the calibration or regression matrix, **P**. **P** contains the calibration, or regression, coefficients which are used to predict the concentrations of an unknown from its spectrum. **P** will contain one row of coefficients for each component being predicted. Each row will have one coefficient for each spectral wavelength. Thus, **P** will have as many columns as there are spectral wavelengths.

Since the data in **C** and **A** contain noise, there will, in general, be no exact solution for equation [46], so, we must find the best least-squares solution. In other words, we want to find **P** such that the sum of the squares of the errors is minimized. The errors are the difference between the measured concentrations, **C**, and the concentrations calculated by multiplying **P** and **A**:

$$\text{errors} = \mathbf{P\,A} - \mathbf{C} \qquad [48]$$

To solve for **P**, we first post-multiply each side of the equation by $\mathbf{A^T}$, the transpose of the absorbance matrix.

$$\mathbf{C\,A^T} = \mathbf{P\,A\,A^T} \qquad [49]$$

Recall that the matrix $\mathbf{A^T}$ is formed by taking every row of **A** and placing it as a column in $\mathbf{A^T}$. Next, we eliminate the quantity $[\mathbf{A\,A^T}]$ from the right-hand side of equation [49]. We can do this by post-multiplying each side of the equation by $[\mathbf{A\,A^T}]^{-1}$, the matrix inverse of $[\mathbf{A\,A^T}]$.

$$\mathbf{C\,A^T}[\mathbf{A\,A^T}]^{-1} = \mathbf{P}\,[\mathbf{A\,A^T}]\,[\mathbf{A\,A^T}]^{-1} \qquad [50]$$

$[\mathbf{A\,A^T}]^{-1}$ is known as the pseudo inverse of **A**. Since the product of a matrix and its inverse is the identity matrix, $[\mathbf{A\,A^T}]\,[\mathbf{A\,A^T}]^{-1}$ disappears from the right-hand side of equation [50] leaving

$$\mathbf{C\,A^T}[\mathbf{A\,A^T}]^{-1} = \mathbf{P}\,[\mathbf{A\,A^T}] \qquad [51]$$

In order for the inverse of $[\mathbf{A\,A^T}]$ to exist, **A** must have at least as many columns as rows. Since **A** has one row for each component and one column for each wavelength, this means that we must have at least as many samples as wavelengths in order to be able to compute equation [51]. In our case, we have spectra of 100 wavelengths each, but only 15 samples in our training set. Obviously we have a problem here. If we were working with actual spectra measured at each of 300, or 1500 wavelenths, it would not generally be practical to assemble enough samples to use ILS. For equation [51] to be meaningful, we also must have at least as many wavelengths as there are components.

It is because of this requirement for at least as many samples as wavelengths that CLS is often called a (or the) *whole spectrum method* to contrast it with ILS. In fact the term is frequently employed in a derogatory way with respect to ILS to suggest that ILS is, by contrast to CLS, not a whole spectrum method.

We have two ways to proceed. We can either select up to 15 individual wavelengths from our 100 wavelength spectra, or we can search for a way to condense these 100 wavelengths into some smaller set of numbers. There are many examples in the literature of the first approach. These examples are often part of a publication that purports to show that ILS does not, in general, produce calibrations that perform as well as those produced by CLS. Usually calibrations developed with a relatively small number of wavelengths that are selected in a casual way will generally not perform as well as a calibration developed with the benefit of signal averaging the noise of 100 or 1000 individual wavelengths. On the other hand, calibrations developed with optimally selected wavelengths can perform extremely well.

It is important to note that there are many publications which discuss optimal ways of selecting individual spectral wavelengths for use with ILS. Much of this work comes from the near infrared (NIR) community. It provides many examples of the power of intellegent wavelength selection. Unfortunately these methods often require more computional time and power than is convenient.

We can also condense the dimensionality of our spectra in other ways. One of the most common, and often one of the best, ways is to work with integrated areas of analytically important spectral peaks. We will see in the next chapters, that the factor based methods, PCR and PLS, are nothing more than ILS conducted on data that is first optimally compressed.

Predicting Unknowns

Now that we have calculated **P,** we can use it to predict the concentrations in an unknown sample from its measured spectrum. First, we place the spectrum into a new absorbance matrix, A_{unk}. We can now use equation [46] to give us a new concentration matrix, C_{unk}, containing the predicted concentration values for the unknown sample. Notice the similarity to equation [39].

$$C_{unk} = P \ A_{unk} \qquad [52]$$

Notice the similarity of equation [52] to equation [39].

Condensing the Data

Our synthetic data simulate spectra that are measured at 100 discrete wavelengths. But, we only have 15 spectra in our training set. Thus, before we can perform ILS on our data, we must first condense our training set data to no more than 15 individual data points per spectrum. Another way of saying this is that the spectral data has an original dimensionality of 100, and we have to reduce the dimensionality to 15 or less. At this point, we would normally define some number of analytical regions over which to integrate the spectra. Our choice of regions would be based upon what we know about the samples and the spectral activity of their components. If this selection of regions is performed well, it will usually lead to a calibration that works well. Unfortunately, this process can be very labor intensive. For our purposes, we will not worry about finding an optimum set of analytical regions. In fact, we will use a basic, simple-minded approach that we would expect, in general, is not optimum. We will simply sum our spectra into 10 "bins" of 10 wavelengths each. For each spectrum, the absorbances at wavelengths 1 through 10 will be added together and placed into the first "bin." The sum of the absorbances at wavelengths 11 through 20 will be placed into the second bin, and we will continue in this fashion until all the wavelengths have been summed into their respective bins. The result, for each data set, is a set of condensed (and possibly degraded) spectra of 10 data points each. Since we have 15 samples in each training set, we will be able to apply ILS to these condensed spectra. Figure 27 contains plots of the condensed spectra we will use in place of the original spectra in **A1** through **A5**. The matrices containing the condensed spectra are named **AA1** through **AA5**, respectively.

These condensed spectra certainly don't look like much. We can see that many of the spectra in **AA4** have higher absorbances than the two training sets. Also, some extraneous looking peaks can be seen in **AA5**.

ILS Results

We now can subject the condensed data to ILS. We will generate two calibration matrices: **P1** from the condensed spectra training set with the structured design, **AA1** and **C1**, and **P2** from the training set with the random design **AA2** and **C2**. We will then use these calibrations to predict the concentrations from all the sets of condensed spectra **AA1** through **AA5**. Following our naming convention, we will assemble the results in matrices named **P1**$_{res}$ through **P25**$_{res}$. **P1**$_{res}$ will hold the concentrations for the training set **AA1** predicted by **P1**, the calibration generated by that training set. **P24**$_{res}$ will hold the concentrations predicted for validation set **AA4** using calibration **P2**, etc. Remember, all of these names are summarized in the crib sheet inside the back cover. Figure 28 contains plots of the expected vs. the predicted concentrations for all of the data sets. Table 5 contains the values for PRESS, SEC^2, SEP^2, and r, for this set of results.

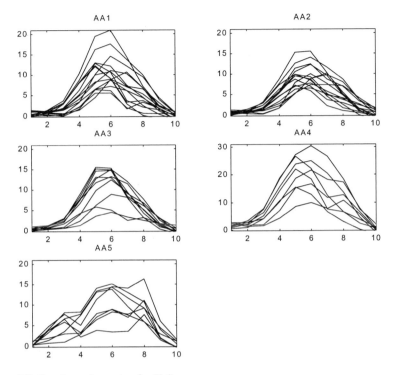

Figure 27. Condensed spectra for ILS.

	P1			P2		
	PRESS	SEC^2	r	PRESS	SEC^2	r
AA1	.0007	.0024	.9980	-	-	-
AA2	-	-	-	.0003	.0011	.9983
AA3	.0057	.0052	.9720	.0043	.0044	.9803
AA4	.0043	.0045	.9961	.0136	.0124	.9960
AA5	.2537	.2603	.8341	.2397	.2043	.8238

Table 5. PRESS, SEC^2, SEP^2, and r for **P1$_{res}$** through **P25$_{res}$**.

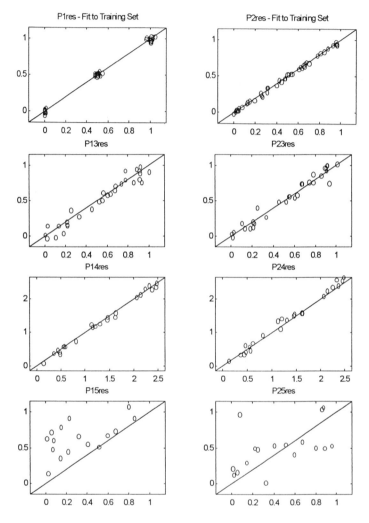

Figure 28. Expected concentrations (x-axis) vs. predicted concentrations (y-axis) for nonzero intercept ILS calibrations (see text).

We can see that the ILS calibrations are noticeably better than the CLS calibrations done with zero intercept. And they are as good or somewhat better that the CLS calibrations with nonzero intercept. This is remarkable when we consider how badly we have degraded the spectra when we condensed them! The main reason for the advantage of ILS over CLS can be seen in equation [47].

Unlike CLS, ILS does not require that we provide concentration values for *all* of the components present. In equation [47], we are not trying to account for all of the absorbances in the spectra. Instead, the formulation allows us to pick up only that portion of the spectral absorbance that correlates well to the concentrations. If ILS is able to do so well with degraded spectra, imagine how much better we might do if we can find a more optimum way of reducing the dimensionality of the spectra than simply summing them into bins. That is precisely what PCR and PLS will do for us.

Every lady in this land
Has 20 nails upon each hand
5 and 20 on hands and feet
All this is true without deceit.
— Auld English Rime

Factor Spaces

We are about to enter what is, to many, a mysterious world—the world of factor spaces and the factor based techniques, Principal Component Analysis (PCA, sometimes known as Factor Analysis) and Partial Least-Squares (PLS) in latent variables. Our goal here is to thoroughly explore these topics using a data-centric approach to dispell the mysteries. When you complete this chapter, neither factor spaces nor the rhyme at the top of this page will be mysterious any longer. As we will see, it's all in your point of view.

Eliminating the Barriers

Many analytical practitioners encounter a serious mental block when attempting to deal with factor spaces. The basis of the mental block is twofold. First, all this talk about abstract vector spaces, eigenvectors, regressions on projections of data onto abstract factors, etc., is like a completely alien language. Even worse, the techniques are usually presented as a series of mathematical equations from a statistician's or mathematician's point of view. All of this serves to separate the (un?)willing student from a solid relationship with his data; a relationship that, usually, is based on visualization. Second, it is often not clear why we would go through all of the trouble in the first place. How can all of these "abstract", nonintuitive manipulations of our data provide any worthwhile benefits?

Our first task is to knock this barrier down to size. Instead of facing a brick wall, we should simply become aware that we are about to cross over a threshold, and a low threshold, at that. This author feels uniquely qualified to guide you over this threshold, having, himself, become a chemist because of a pronounced aversion to heavy mathematics.

The first step in our journey is to realize that there is nothing so unusual about these factor-based techniques. In fact, it is likely that you have already used one or more factor spaces in your studies or your work without even realizing it! You see, a factor space is nothing more than a particular coordinate system that offers certain advantages to the task at hand. When we operate in a factor space, instead of the native data space, we are simply mapping our data into a new coordinate system. We are not actually changing the data itself. The operation is no more difficult or mysterious that converting from rectangular to polar coordinates, and back again. Some factor spaces you might already have used are:

1. **Polar Coordinates**. Yes, this is a type of factor space. We might use polar coordinates when we are mapping electron densities, seasonal population variations, or anytime it makes our work more convenient or allows us to better visualize or understand our data.

2. **Fourier Series**. Most of us are quite comfortable with the concept (if not the mathematics) of mapping a signal or other data back and forth between the time domain to the frequency domain. When we take a time domain signal and represent it as a combination of cosine waves, this is nothing more than a transformation of our coordinate system. The coordinates of the signal are changed from time and amplitude to frequency and amplitude. The signal, itself, is unchanged.

3. **Taylor Series**. Most of us have practiced how to approximate a curve over a bounded region as a series of power terms: $y = a_0 + a_1x + a_2x^2$ ... . But, we probably never realized that each power term x, x^2, ..., can be considered as a new coordinate axis, and each coefficient a_0, a_1, ..., is simply the new coordinate on its respective axis.

4. **Eigenvectors and PCA**. This is the factor space we are about to explore. We will be working with a factor space defined by the eigenvectors of our data simply because a coordinate system of eigenvectors has certain properties that are convenient and valuable to us.

5. **PLS**. Also on our agenda. We will soon see that PLS is simply a variation of PCA.

Very often, the axes of the new coordinate system, or factor space are chosen to be mutually orthogonal, but this is not an absolute requirement. Of the above examples, the axes chosen for 3 and 5 are generally not mutually orthogonal.

There are several reasons why we might want to use the coordinate system of a factor space rather than the native space comprised of physically meaningful coordinates:

1. **Numerical conditioning**. By mapping our data from the native coordinate system into an appropriate factor space, we can eliminate problems caused by highly colinear data such as a set of very similar spectra. This can reduce calculation round-off errors and make it possible

to perform calibrations that are difficult or impossible in the normal coordinate space.

2. **Reduced assumptions.** By using an appropriate factor space, we may be able to eliminate some assumptions about our data that aren't always completely true. Examples of assumptions that can hurt us are: linearity of the data, independence of samples, number of components.

3. **Noise rejection.** Factor spaces can provide a superior way of removing noise from our data.

4. **New insights into the data.** Mapping our data into an appropriate factor space can provide a new frame of reference wherein patterns that were not apparent in the native coordinate space become evident. Factor spaces can help us understand how many components are actually present, or which samples are similar or different to which other samples.

5. **Data compression.** As mentioned in the previous chapter, PCA and PLS provide us with an optimum way to reduce the dimensionality of our data so that we can use ILS to develop calibrations.

Visualizing Multivariate Data

We have already touched on the need to visualize our multivariate data in a multivariate way. We will now consider this topic in more detail. Earlier, we mentioned that Figure 1 contains the most important concepts in this entire book. The next series of figures are the second most important. Once you understand the concepts in the next few figures, you will be well on your way to mastering the factor-based techniques.

Let's consider a hypothetical set of spectral measurements on a two component system. We will measure our spectral absorbances at three separate wavelengths. Each of the two components absorbs a different amount of light at each of the three wavelengths. To make things simple, let's start with the case were there is no noise, no baseline drift, and no nonlinearities. In other words, we will discuss the perfectly linear, noise-free case. Referring to Figure 29, we recall that we can plot the spectrum of any mixture of these two components as a unique point in a 3-dimensional space. The absorbance at each wavelength of the 3-wavelength spectrum is plotted along a separate axis.

First, consider a blank sample in which the concentration of both components is equal to zero. With no absorbing species in the sample, there would be no absorbance at any of the wavelengths, and the spectrum would be plotted at [0, 0, 0], the origin of this absorbance data space. Now, consider the spectrum of a sample that contains 1 concentration unit of Component 1 and none of Component 2. This spectrum will have some absorbance at each of the

3 wavelengths. We can plot this spectrum in the 3-dimensional space by plotting the absorbance of the first wavelength along one axis, the absorbance of the second wavelength along, another axis, and the absorbance of the third wavelength along the remaining axis as shown in Figure 29.

Next, let's consider the spectrum of a sample that contains 2 concentration units of Component 1 and none of Component 2. In this perfectly linear, noise-free case, when we double the concentration of Component 1, the absorbance at each of the wavelengths will also double. We have also plotted this second spectrum in Figure 29. It lies along the same direction from the origin as the first spectrum and at twice the distance.

In similar fashion, we have plotted the spectrum of a third sample which contains 3 concentration units of Component 1 and none of Component 2. Of course, this spectrum also lies in the same direction from the origin as the first spectrum and at 3 times the distance. It is clear that, when we use this approach to plot the spectra of samples which contain only Component 1, each such spectrum must lie somewhere along a line extending from the origin of the data space in some unique direction that is determined by the relative absorbances of Component 1 at each of the wavelengths plotted.

Now let's consider a sample that has 1 concentration unit of Component 2 and none of Component 1. Since the spectrum of Component 2 is different from that of Component 1, this sample has absorbances at each of the 3 wavelengths that are different from those of the samples plotted in Figure 29. The spectrum

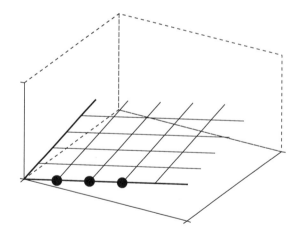

Figure 29. Multivariate plot of 3-wavelength spectra for samples containing only Component 1.

for this sample is plotted in Figure 30. Notice that, since the spectrum for this sample is different from the spectrum of Component 1, it must lie in some other direction from the origin and at some other distance. We should also realize that the angle between this sample and the first three samples is totally arbitraty. It will depend solely on the nature of the differences in the spectra of the two components. We can continue to plot spectra of samples which contain only different concentrations of Component 2, just as we did for the samples that contained only Component 1. A total of three such samples are plotted in Figure 30.

Next, let's consider the spectrum of a sample that contains both components together: 2 concentration units of Component 1 and 3 concentration units of Component 2. Figure 31 contains a plot of this sample. The heavy X^{s} are plotted to indicate the absorbance contribution from each of the pure components in the sample. Since the contribution to the absorbance at each wavelength from each component adds linearly, the spectrum of the mixture is identical to the vector addition of the spectra of the pure components. Thus, it is apparent that, if we were to plot the spectrum of any mixture of these two components, it must be located somewhere in the plane determined by the lines which lie along the directions of the two pure component spectra. Notice that these lines that define the plane do not have to be perpendicular to eachother. Indeed, they will usually not be mutually orthogonal. Figure 32 shows a plot of a number of such samples for this noise-free, perfectly linear case.

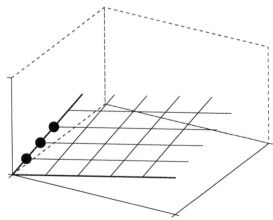

Figure 30. Multivariate plot of 3-wavelength spectra for samples containing only Component 2.

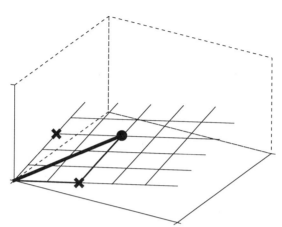

Figure 31. Multivariate plot of 3-wavelength spectra for samples containing both Component 1 and Component 2.

We can easily see that, even though each spectrum was measured at three wavelengths, all of the spectra in Figure 32 lie in a plane. Suppose that instead of measuring at only three wavelengths, we measured each spectrum at 100 different wavelengths. If we were able to plot these 100-dimensional spectra on a two dimensional page, we would see that all of the spectra would lie in a 2-dimensional plane oriented at some angle in the 100-dimensional hyperspace.

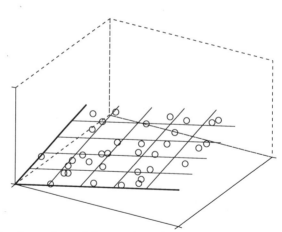

Figure 32. Multivariate plot of 3-wavelength spectra for samples containing varying amounts of Component 1 and Component 2. In this noise-free, linear case, all of the specra must lie in a plane.

If we could find a pair of axes that lay in this plane, we could use these axes as the basis of a new coordinate system. We could then simply specify each spectrum in terms of its distance along each of the two axes of our new coordinate system. Notice that this doesn't change the data at all. The data points do not move when we change coordinate systems. This is no different than deciding to define a point in space in terms of its polar coordinates rather than its rectangular coordinates. We would no longer need to provide 100 individual numbers to identify each spectrum by its 100-dimensional spectrum. We could, instead specify each spectrum of this 2 component system by just two numbers, the distances along each new coordinate axis. By extension, the spectra of a 3 component system would require three numbers, a 4 component system 4 numbers, etc.

Welcome to Our Abstract Factor Space

What does all of this have to do with factor spaces? Any pair of axes lying in the plane which holds the spectra comprise a factor space for that data. Each axis is a factor of the data space. These are usually called *abstract factors* because they usually do not have an easily interpretable physical meaning. Instead of specifying each spectrum in terms of its 100 wavelengths, we would specify each spectrum in terms of its projections onto the factors. These projections are often called the *scores*. Note that the projections of a spectrum onto the factors is nothing more than the distance of that spectrum along the direction of each factor. If we have two factors, as we do in this case, we get two projections, one for each factor. Thus, we can compress the 100-dimension spectra into 2 dimensions *without any distortion of the data*. The data remain unharmed, because none of the data points has been changed in any way. We have simply found a more efficient way to express the data. So we have seen that the concept of "working with the projections of our data onto an abstract factor space" is nothing more than a long, obfuscative way of saying that we are using a more convenient coordinate system.

So, we have crossed into the formerly mysterious world of factor spaces. In doing so, we have discovered that the barrier to entry was not a brick wall, but merely a threshold after all.

Finding the Factors

Now let's consider how we find a pair of axes that lie in the plane containing these data. Let's not aim to find just any two axes. It would be convenient to have factors that are mutually orthogonal, so that changes along one axis do not interact with changes along the second axis. Let's also try to find the set of factors that span our data as efficiently as possible. As we will soon see, insisting that each factor is as efficient as possible is key to their utility.

Beginning with the data in Figure 32, let's find the single axis that most efficiently spans the data points. In other words, we are looking for the axis that captures as much of the variation, or variance, in the data as it possibly can. Yet another way of saying this is that we want to find the unique axis in this 3-dimensional absorbance space for which the sum of the squares of the distances of all the data points from that axis is a minimum. It turns out that this axis is exactly the same thing as the first *eigenvector* for this set of data. In other words, if we put all of the spectra into an absorbance matrix, and calculate the first eigenvector for that matrix, that eigenvector will be such that the squares of all the distances between the vector and all of the data points will be the minimum possible. This means that this first eigenvector spans the maximum variance of the data that can be spanned with a single vector. That is why this eigenvector is also called the first *principal component* of this data set. Figure 33 contains a plot of the first eigenvector for the data. Figure 34 contains different views of the same plot, one view edge on to the plane containing the points, and a second view looking down onto the plane perpendicularly. These views make it easy to see that the eigenvector lies in the plane containing the data. This eigenvector *must* lie in the plane of the data points if it is to span the maximum possible variance in the data. Stated another way, any motion of the eigenvector above or below the plane of the data, will increase the sum of the squares of the distances of the points to the vector.

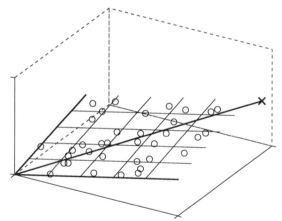

Figure 33. The data from Figure 32 plotted together with the first eigenvector (factor) for the data.

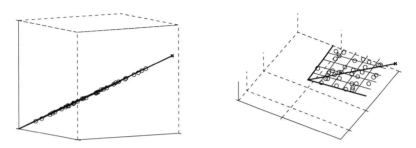

Figure 34. The data from Figure 32 plotted together with the first eigenvector (factor) for the data.

Now let's find the next vector that spans the maximum possible amount of the remaining variance that was not spanned by the first factor. It turns out that this vector is identical to the second eigenvector of the data. This vector must be orthogonal to the first factor. If it were not orthogonal, it could not capture the maximum amount of the remaining variance. It must also lie in the plane of the data for the same reason. Figure 35 contains a plot of the data together with the first two factors. Figure 36 shows two different views that make it easy to see that both factors lie in the plane of the data and are perpendicular to each other.

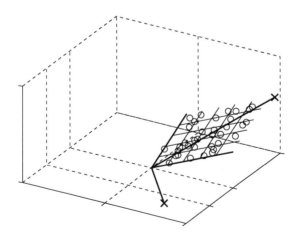

Figure 35. The data from Figure 32 plotted together with the first two eigenvectors (factors) for the data.

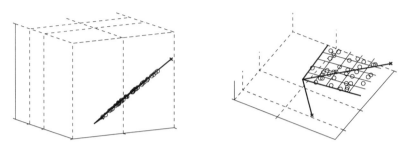

Figure 36. The data from Figure 32 plotted together with the first two eigenvectors (factors) for the data.

So we have found a pair of axes that we can use as the basis of a new coordinate system. And since each axis spans the maximum possible amount of variance in the data, we can be assured that there are no axes that can serve as a more efficient frame of reference than these two. Each axis is a factor or principal component of the data. Together, they comprise the *basis space* of this data set.

Even though two factors are all we need to span this data, we could find as many factors as there are wavelengths in the spectra. Each successive factor is identical to each successive eigenvector of the data. Each successive factor will capture the maximum variance of the data that was not yet spanned by the earlier factors. Each successive factor must be mutual orthogonal to all the factors that precede it. Let's continue on and plot the third factor for this data set. The plots are shown in Figures 37 and 38.

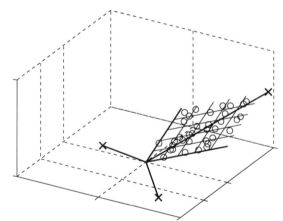

Figure 37. The data from Figure 32 plotted together with all 3 eigenvectors (factors) for the data.

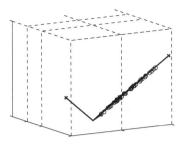 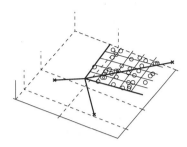

Figure 38. The data from Figure 32 plotted together with all 3 eigenvectors (factors) for the data.

We can see that the third factor is, indeed, orthogonal to the previous two. In these plots, each factor has been plotted at a different length for clarity. In reality, since each vector merely serves to define the direction of a coordinate axis, its length, or magnitude, is irrelevant. The length of the vectors is typically normalized to unity. Similarly, the sign of the vector is completely arbitrary. In fact, different algorithms for calculating the eigenvalues of a data set will produce the same eigenvectors, but the signs of the individual vectors will often be different.

Eigenvalues

Each eigenvector has an eigenvalue associated with it. The eigenvalue of a eigenvector is equal to the sum of the squares of the projections of the data onto the eigenvector. Remember, the projections are nothing more than the distance along the vector of each data point. The eigenvalue is, thus, a measure of the total variance captured, or spanned, by the eigenvector. Table 6 contains the eigenvalues for the 3 eigenvectors of the data we have been considering.

Eigenvector #	Eigenvalue
1	30.5390
2	1.7298

Table 6. Eigenvalues of the eigenvectors for the data in Figure 32.

We notice that the eigenvalue of each successive eigenvector is less than that of its predecessor. This makes sense because each eigenvector captures the maximum possible variance it can. Each succeeding eigenvector is capturing the variance in the residuals that are left behind by all of its predecessors. Since the residuals must get smaller and smaller, each successive eigenvector has less and less variance available for capture, so each successive eigenvalue must be smaller than the ones preceeding it. We also notice that the eigenvalue of the last eigenvector is exactly zero. This also makes sense because, for this noise-free, prefectly linear case, all of the data lie precisely in the plane defined by the first two eigenvectors. Since the third eigenvector must be orthogonal to the other two, it must be orthogonal to the plane holding the data. Thus, the projection of each data point onto the third eigenvector must be exactly zero.

Data with Noise

Now we are ready to consider what happens if the data are noisy. We will take the data we just used and add some noise to it. We will add normally distributed noise to each wavelength of each spectrum at a level of approximately 5%. It is important to understand that, within a given spectrum, the particular amount of noise added to each wavelength is independent of the noise added to the other wavelengths. And, of course, the noise we add to each spectrum is independent of the noise added to the other spectra. In other words, there is no correlation to the noise. Figure 39 contains a plot of the data before and after the addition of the noise. Figure 40 show two other views of the data after the additon of the noise.

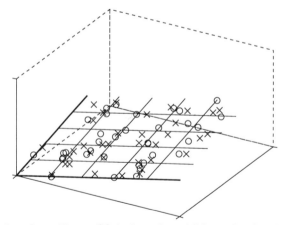

Figure 39. The data from Figure 32 before the addition of noise (x) and after the addition of noise (o).

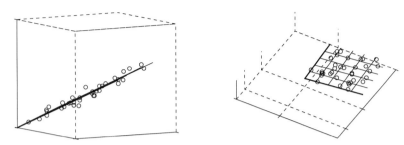

Figure 40. The data from Figure 32 after the addition of noise.

We can see that the effect of the noise has been to displace each data point from its original location. Since, for each spectrum, the amount of noise at each wavelength is independent of the noise at the other wavelengths, each data point is moved a different, randomly determined distance, in a different, randomly determined direction. Since the direction and distance each data point moved is totally random, we can say that the noise is *isotropic,* that is, uniform in all directions of this data space. While the displacements have a component within the original plane of the data, they also have a component perpendicular to the original plane of the data. Thus, we see in Figure 40, that the points no longer all lie perfectly within the original plane.

Next, we find the first eigenvector of the noisy data set and plot it in Figures 41 and 42. We see that it is nearly identical to the first eigenvector of the noise

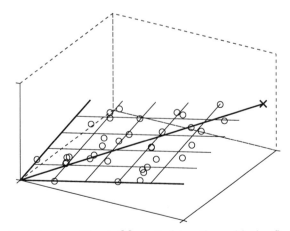

Figure 41. The noisy data from Figure 39 plotted together with the first eigenvector (factor) for the data.

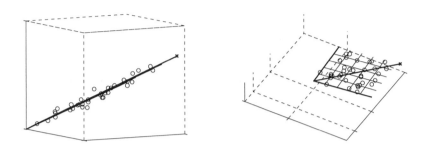

Figure 42. The noisy data from Figure 39 plotted together with the first eigenvector (factor) for the data.

free data. If we look extremely closely, we can detect a very slight displacement as compared to the noise-free eigenvector. This makes sense when we realize that, in capturing the maximum possible variance in the data by minimizing the sum of the squares of the distances from the vector to the data points, the vector has simultaneously performed a least-squares average of all the noise in the data.

Continuing, we find the second eigenvector for the noisy data. Figures 43 and 44 contain plots of the first two eigenvectors for the noisy data. Again, the second eigenvector for the noisy data is nearly identical to that of the noise-free data.

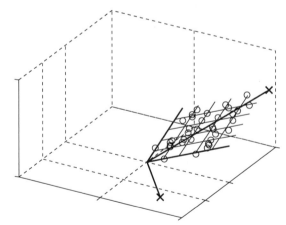

Figure 43. The noisy data from Figure 39 plotted together with the first two eigenvectors (factors) for the data.

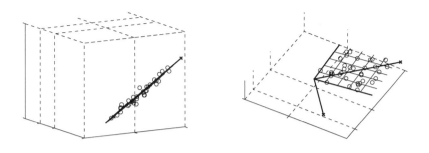

Figure 44. The noisy data from Figure 39 plotted together with the first two eigenvectors (factors) for the data.

Completing the cycle, we calculate the third eigenvector for the noisy data. Figures 45 and 46 contain the plots of all three eigenvectors for the noisy data.

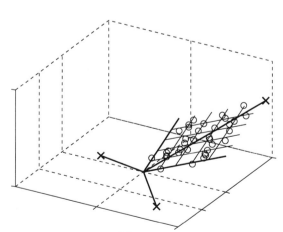

Figure 45. The noisy data from Figure 39 plotted together with all three eigenvectors (factors) for the data.

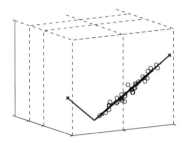 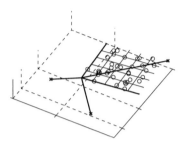

Figure 46. The noisy data from Figure 39 plotted together with all three eigenvectors (factors) for the data.

Finally, we calculate the eigenvalues for these eigenvectors. They are show in Table 7 together with the eigenvalues for the noise free data.

Discarding Some Noise

Referring to Table 7, we see that the eigenvalue for the third eigenvector of the noisy data is no longer equal to zero. Of course, this makes perfect sense because the noisy data no longer lie exactly in a plane and so the third eigenvector is now able to capture some variance from the data.

Let's consider the nature of the variance spanned by the third eigenvector. We know that it cannot contain any information that is related to the concentrations of the components in the samples because that information can only lie in the plane of the original data. Thus, the Information-to-Noise ratio of the variance spanned by this eigenvector must be zero.

Since, we made this data ourselves, we know with certainty that it contains only two components (and it is perfectly linear). So we know that, no matter how many individual wavelengths we decide to use when we measure the

Eigenvector #	Noise-free Eigenvalue	Noisy Eigenvalue
1	30.5390	30.5880
2	1.7298	1.6394

Table 7. Eigenvalues of the eigenvectors for the data in Figures 32 and 39.

spectra, the information in those spectra must all lie in some plane in the n-dimensional spectral space. Therefore, we can confidently decide to discard that portion of the variance in the data that displaces the data points out of the plane. We will soon see that there are ways to determine how many vectors to discard in the absence of the kind of *a priori* information we have in this case.

So, we can discard the third eigenvector and, along with it, that portion of the variance in our spectra that displaced the data out of the plane of the noise-free data. We are in fact, discarding a portion of the noise without significantly distorting the spectra! The portion of the noise we discard is called the *extracted error* or the *residuals*. Remember that the noise we added also displaced the points to some extent within the plane of the noise-free data. This portion of the noise remains in the data because it is spanned by the eigenvectors that we must retain. The noise that remains is called the *imbedded error*. The total error is sometimes called the *real error*. The relationship among the real error (RE), the extracted error (XE), and the imbedded error (IE) is

$$RE^2 = IE^2 + XE^2 \qquad [53]$$

The eigenvectors that we discard are sometimes called the *error eigenvectors, noise eigenvectors*, or *secondary eigenvectors*. The eigenvectors we keep are called the *basis vectors*, *principal components*, *loadings*, *primary eigenvectors*, or *factors*, of the data. The number of factors that we retain is called the *rank*, or the *dimensionality*, of this subset of the data. It is important to remember that we are assuming here that we are correctly discarding all the vectors that model only noise.

Since the noise is isotropic, each vector, whether a noise vector or a basis vector, picks up its equivalent share of the noise (we will see, soon, that we should take degrees-of-freedom into account when discussing what amount of noise is an equivalent share for each vector). If we had measured the spectra of our 2-component system at 100 wavelengths, we would, potentially be able to discard 98 out of a possible 100 eigenvectors. In doing so, we would expect to discard more noise than we can in this case.

This process of discarding the noise eigenvectors to extract some of the noise from the data is sometimes called *short circuit data reproduction*. A more convenient term is *regeneration*.

Let's see what our data look like when we regenerate it after discarding the variance spanned by the third eigenvector. Figure 47 contains a plot of the regenerated data and the original, noise-free data. If we compare this plot to the plot in Figure 39, we see that the regenerated points, as a whole, lie closer to the noise-free points than do the original noisy points. We can also see this numerically. If we calculate the sum of the squares of the differences between the noise-free data and the noisy data we get 0.2263. The same calculation with the regenerated data yields 0.1623.

Figure 48 is an edge-on view of the plane of the original noise-free data together with the regenerated data. We can see that the regenerated data lie exactly in a plane, but this plane is not precisely in line with the plane of the original, noise-free data. This is because, the noise in the data deflected the first eigenvector of the noisy data slightly above the plane of the noise-free data.

So now we understand that when we use eigenvectors to define an "abstract factor space that spans the data," we aren't changing the data at all, we are simply finding a more convenient coordinate system. We can then exploit the properties of eigenvectors both to remove noise from our data without significantly distorting it, and to compress the dimensionality of our data without compromising the information content.

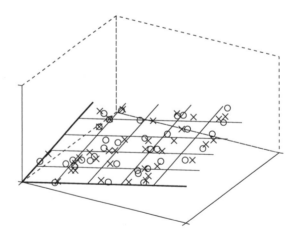

Figure 47. The noise-free data from Figure 32 (x) plotted with the data from Figure 39 as regenerated with only the first two factors (o).

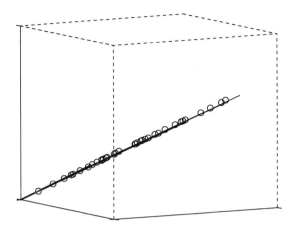

Figure 48. The plane of the original noise-free data viewed edge on together with the data from Figure 39 as regenerated with only the first two factors.

Factor spaces are a mystery no more! We now understand that eigenvectors simply provide us with an optimal way to reduce the dimensionality of our spectra without degrading them. We've seen that, in the process, our data are unchanged except for the beneficial removal of some noise. Now, we are ready to use this technique on our realistic simulated data. PCA will serve as a pre-processing step prior to ILS. The combination of Principal Component Analysis with ILS is called Principal Component Regression, or PCR.

Every lady in this land
Has 20 nails, upon each hand
5, and 20 on hands and feet.
All this is true without deceit.
— **Auld English Rime**

Principal Component Regression

Principal component regression is sometimes described as "performing a least-squares regression of the projections of the data onto the basis vectors of a factor space using ILS." We have seen, in the previous chapter, that this is just a long and obscure way to say that we are generating an ordinary ILS calibration but using a different coordinate system to specify our spectra. PCR is a multistep operation. Figure 1 contains a flow chart showing the steps.

Recall that, in order to generate an ILS calibration, we must have at least as many samples as there are wavelengths used in the calibration. Since we only have 15 spectra in our training sets but each spectrum contains 100 wavelengths, we were forced to find a way to reduce the dimensionality of our spectra to 15 or less. We have seen that principal component analysis (PCA) provides us with a way of optimally reducing the dimensionality of our data without degrading it, and with the added benefit of removing some noise.

Optional Pretreatment

Even though we have waited until this point to discuss optional pretreatments, they are equally applicable to CLS, ILS, PCR, and PLS. There are a number of possible ways to pretreat our data before we find the principal components and perform the regression. They fall into 3 main categories:

1. artifact removal and/or linearization
2. centering
3. scaling and weighting

Optional pretreatments can be applied, in any combination, to either the spectra (the x-data), the concentrations (the y-data) or both.

Artifact removal and/or **linearization**. A common form of artifact removal is baseline correction of a spectrum or chromatogram. Common linearizations are the conversion of spectral transmittance into spectral absorbance and the multiplicative scatter correction for diffuse reflectance spectra. We must be very careful when attempting to remove artifacts. If we do not remove them correctly, we can actually introduce other artifacts that are worse than the ones we are trying to remove. But, for every artifact that we can correctly remove from the data, we make available additional degrees-of-freedom that the model can use to fit the relationship between the concentrations and the absorbances. This translates into greater precision and robustness of the calibration. Thus, if we can do it properly, it is always better to remove an artifact than to rely on the calibration to fit it. Similar reasoning applies to data linearization.

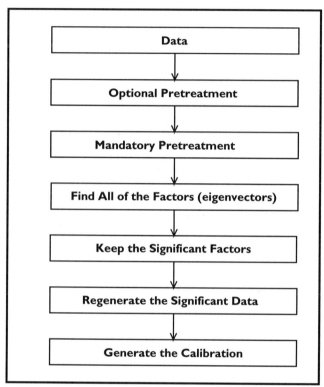

Figure 49. The steps of a PCR calibration.

Centering, sometimes called *mean centering*, is simply the subtraction of the mean absorbance at each wavelength from each spectrum. In other words, we compute the mean spectrum for the data set and subtract it from each spectrum. This shifts the origin of our coordinate system to the center of the data set. Interestingly enough, in some circles it is controversial to suggest that centering is an *optional* step. Because of this, you might see some published work where data centering was employed, but the author, believing that centering is mandatory and routine, doesn't even see fit to mention the fact. The worst part of the situation is that some software packages do not allow you to choose whether or not to center the data. The main reason for centering data is to prevent data points that are farther from the orgin form exerting an undue amount of leverage over the points that are closer to the origin. While there are sound, statistically-based arguments in favor of this practice, they are based on assumptions that often do not apply to real chemical data. If we center our data, we lose information about the origin of the factor space. We also lose information about the relative magnitudes of the eigenvalues, and the relative errors. Depending upon the data and the application, this can have undesireable consequences. Additional discussion of centering can be found in Appendix C.

Scaling and **weighting**. There are many possible ways to scale or weight our data. Scaling or weighting involves multiplying all of the spectra by a different scaling factor for each wavelength. This is done to increase or decrease the influence on the calibration of each particular wavelength. The most basic form of weighting is to select which spectral wavelengths to include or exclude from the calibration—the included wavelengths are scaled by a factor of 1 while the excluded wavelengths are scaled by a factor of 0. Two types of scaling are commonly encountered, *variance scaling*, which is sometimes called *normalization*, and *autoscaling*. As is the case with data centering, in some circles it is controversial to suggest that data scaling is *optional*. However, *unlike* data centering, data scaling can often be *very* detrimental to the precision and/or robustness of a calibration. Thus, it is particularly onerous that some software packages do not allow you to choose whether or not to scale your data. Additional discussion of centering and weighting can be found in Appendix C.

Mandatory Pretreatment

Whether or not we scale, weight, and/or center our data, a mandatory pretreatment is required by most of the algorithms used to calculate the eigenvectors. Most algorithms require that we square our data matrix, **A**, by either pre- or post-multiplying it by its transpose:

$$\mathbf{D} = \mathbf{A}^T \mathbf{A} \qquad [54]$$

$$\mathbf{D} = \mathbf{A} \mathbf{A}^T \qquad [55]$$

It doesn't matter if we use equation [54] or [55] to square our data matrix—the information in the matrix will be unchanged in either case. But, if we do not have the same number of samples as wavelengths, equations [54] and [55] will produce different sized matrices, **D**. For our training sets which contain 100 wavelengths and 15 spectra, equation [54] will produce **D** with 15 rows and 15 columns while equation [55] will produce **D** with 100 rows and 100 columns. Either matrix, **D**, will give us the exact same eigenvectors (except that some of the signs of the various vectors might be different). If we use the 100 × 100 matrix, we will get 100 eigenvectors, but, since we only have 15 samples, only the first 15 eigenvectors can be meaningful. The remaining 85 are useless. If we use the 15 x 15 matrix, we will only get the first 15 eigenvectors. Obviously, the calculation will require considerably less time if we use the smaller matrix. If we had more samples than wavelengths, the situation would flip-flop. In any case, almost every software package that is available handles this detail automatically. We only consider it here to emphasize the equivalency of the two possible square matrices.

Find All of the Factors

We can calculate all of the factors for our data matrix using a number of different algorithms. The two most common are the NIPALS (nonlinear iterative partial least squares) algorithm, and SVD (singular value decomposition). Note that, strictly speaking, we do not generally need to calculate *all* of the factors. We need only calculate the first N factors where N is large enough to enable us to determine how many factors we should include in the basis space.

NIPALS is an iterative algorithm. As such it can suffer from problems with digital round-off error when handling very large data matrices, or data matrices that have a high degree of collinearity. This is the algorithm most commonly referenced in the literature, but it is not necessarily the best algorithm to use. For very large data sets, NIPALS can provide an advantage in that it can easily be stopped after finding the first N factors.

SVD is a way of decomposing a data matrix into factors in a more general sense than NIPALS. We can think of the eigenvectors and eigenvalues of a data matrix as a particular subset of the SVD factors. Most SVD algorithms employ a form of diagonalization that allow for proper management of the scale of the numbers. This helps minimize problems of digital round-off error. As a result, a good SVD algorithm will usually be able to handle even difficult data that can cause a NIPALS algorithm to "blow up."

No matter how they are calculated, the eigenvectors are organized into a matrix which we will call **Vc**. (We might have called the matrix **V**, but **V** is often used to name a particular matrix in singular value decomposition, so we are using a distinct name in order to eliminate the possibility of confusion. Simply understand that **Vc** is the name of a matrix, **c** is not a subscript of a matrix named **V**.) **Vc** is a matrix of column vectors. Each column of **Vc** is an eigenvector, or factor, of the data matrix. **Vc** has as many rows as there are wavelengths in the original spectra. Thus, each eigenvector in **Vc** has an absorbance value for each wavelength in the original spectral space. This means that we can plot each vector in the original wavelength space just as if it were a spectrum. That is precisely how we plotted the eigenvectors in Figures 33 through 46 in the previous chapter.

The length, or magnitude, of each eigenvector is normalized to unity. Thus, the vector cross product of each eigenvector multiplied with itself should be equal to 1. Also, all of the eigenvectors are mutually orthonormal. This means that the vector cross product of any eigenvector times any other eigenvector must equal 0. We can use these last two properties of eigenvectors to check whether or not our software produced a good set of vectors. The correlation matrix, **Rc**

$$\mathbf{Rc} = \mathbf{Vc^T\ Vc} \qquad\qquad [56]$$

gives us the products of all possible cross-products of each vector in **Vc** with itself and the other vectors. The products of every vector with itself, lie on the diagonal of **Rc**, while all of the various cross-products lie off the diagonal. Thus, if our software produced a good set of vectors, **Rc** should have 1's on the diagonal and 0's everywhere else.

Each eigenvector in **Vc** has a corresponding eigenvalue which we will call Vl_n. It is convenient to collect all of the eigenvalues into a single column vector, **Vl**.

Keep the Significant Factors

This is a very important step. If we decide to retain more factors than we should, we would be retaining some factors that can only add more noise to our data. On the other hand, if we do not keep enough factors, we will be discarding potentially meaningful information that could be necessary for a successful calibration. Usually, we do not have enough information about our data, *a priori*, to decide how many factors we should keep. Fortunately, there are a number of tools to help us make the decision:

1. Indicator functions
2. PRESS for validation data
3. Cross-validation

Indicator functions have the advantage that they can be used on data sets for which no concentration values (y-data) are available. But cross-validation and, especially PRESS, can often provide more reliable guidance.

Indicator functions are based upon an analysis of either the eigenvalues or the errors. Some of them have been derived empirically, while others are statistically based. There are a number of empirical functions based on the eigenvalues such as Malinowski's imbedded error function (IE), Malinowski's IND function, and Brown's FRAC function. While all of these work to a certain extent, they are generally not as reliable as might be desired. There are a number of statistically derived indicators based on an analysis of error in the regenerated data vs. the raw data, such as the root-mean-squared error (RMS). But these are often no better than the empirical functions and have a tendency to suggest the retention of more factors than is optimum. A statistically derived indicator, based on an analysis of the eigenvalues, that has proven quite reliable is the 2-way F-test on reduced eigenvalues (REV's) according to the method of Malinowski. It's worthwhile to take some time to understand the concept of reduced eigenvalues.

When we regard each of our spectra as a unique point in the n-dimensional absorbance space, we can say that the error in our data is isotropic. By this, we mean that the net effect of the errors in a given spectrum is to displace that spectrum some random distance in some random direction in the n-dimensional data space. As a result, when we find the eigenvectors for our data, each eigenvector will span its equivalent share of the error. But recall, we said that we must take degrees-of-freedom into account in order to understand what is meant by equivalent share.

To better understand this, let's create a set of data that only contains random noise. Let's create 100 spectra of 10 wavelengths each. The absorbance value at each wavelength will be a random number selected from a gaussian distribution with a mean of 0 and a standard deviation of 1. In other words, our spectra will consist of pure, normally distributed noise. Figure 50 contains plots of some of these spectra. It is difficult to draw a plot that shows each spectrum as a point in a 100-dimensional space, but we can plot the spectra in a 3-dimensional space using the absorbances at the first 3 wavelengths. That plot is shown in Figure 51.

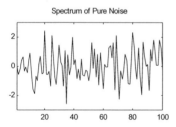 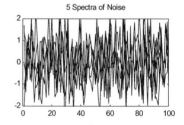

Figure 50. Some spectra consisting of pure, normally distributed noise.

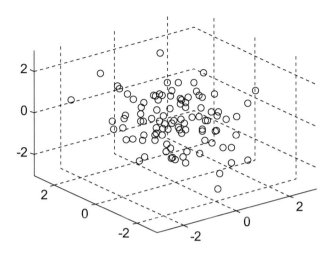

Figure 51. "Absorbances" at the first 3 wavelengths of the spectra containing only noise.

We can see, in Figure 51, that the spectra form a spherical cloud in this 3-dimensional subset of the absorbance data space. In other words, this data is isotropic. No matter in which direction we look, we will see no significant (in the statistical sense of the word) difference in the distribution of the data points. If we were able to show the plot for all 10 dimensions, we would see a 10-dimensional hyperspherical cloud that is isotropic within the spherical distribution of points.

Now let's compute the eigenvectors and eigenvalues of these spectra. We won't attempt to plot the 10-dimensional eigenvectors, but a plot of the eigenvalues is shown is Figure 52. We recall that the eigenvalue for each eigenvector is equal to the amount of variance in the data that is captured by that eigenvector. We can see that the eigenvalues decline steadily from the first one to the last one, over several orders of magnitude. Well, if the data are isotropically distributed, and each eigenvector picks up it's equivalent share of the variance in the data, then why are the eigenvalues not all equal to one another? The reason is that, each time we find another eigenvector, we remove the variance spanned by that eigenvector from the data before we find the next eigenvector. Thus, each time we find another eigenvector we reduce the degrees-of-freedom remaining in the data by 1. The variance remaining in the data is also reduced. Thus, each successive eigenvector is only "entitled" to its equivalent share of the *remaining* variance in the data. Malinowski showed that the remaining variance is proportional to:

$$(w - n + 1)(s - n + 1) \qquad [57]$$

where:

w   is the number of wavelengths in the spectra

s   is the number of spectra, and

n   is the rank (ordinal number) of the eigenvector

Thus, if we wish to compare the eigenvectors to one another, we can divide each one by equation [57] to normalize them. Malinowski named these normalized eigenvectors *reduced eigenvectors*, or REV'^{s}. Figure 52 also contains a plot of the REV'^{s} for this isotropic data. We can see that they are all roughly equal to one another. If there had been actual information present along with the noise, the information content could not, itself, be isotropically distributed. (If the information were isotropically distributed, it would be, by definition, noise.) Thus, the information would be preferentially captured by the earliest

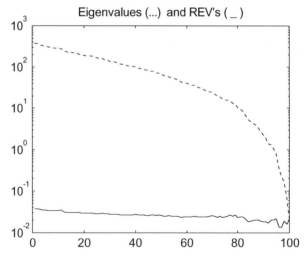

Figure 52. Eigenvalues (...) and reduced eigenvalues (_) for the spectra consisting of pure, normally distributed noise.

eigenvectors until all of the information were spanned. From that point on, the remaining eigenvectors would only span their equivalent share of the residual noise. The eigenvalues of the vectors that spanned information would be significantly (in the statistical sense) larger that the eigenvectors that only spanned noise. We can use the 2-way F-test to ask the question for each eigenvalue: "Is this eigenvalue (statistically) significantly larger than all of the successive eigenvalues?" We begin asking this question with the next-to-last eigenvalue. If the answer is "no," we ask the question again for the eigenvalue immediately preceeding it. We continue in this fashion until the 2-way F-test produces "yes" for an answer. At that point, we retain the eigenvector that goes with that eigenvalue together with all of the eigenvectors preceeding it. These become our basis set—the vectors we retain for our new coordinate system. We will explore the 2-way F-test on REV's in more detail in the next chapter when we perform PCR on our simulated data.

PRESS for validation data. One of the best ways to determine how many factors to use in a PCR calibration is to generate a calibration for every possible rank (number of factors retained) and use each calibration to predict the concentrations for a set of independently measured, independent validation samples. We calculate the predicted residual error sum-of-squares, or PRESS, for each calibration according to equation [24], and choose the calibration that provides the best results. The number of factors used in that calibration is the optimal rank for that system.

Cross-validation. We don't always have a sufficient set of independent validation samples with which to calculate PRESS. In such instances, we can use the original training set to simulate a validation set. This approach is called *cross-validation*. The most commom form of cross-validation is performed as follows:

1. Calculate a calibration matrix using all of the training set samples except for one.
2. Use the calibration to predict the concentrations of the components in the sample that was left out of the training set.
3. Calculate the sum-squared of errors between the expected and predicted concentrations for the sample that was left out.
4. Return the excluded sample to the training set, and leave out a different sample.
5. Calculate a new calibration for this new subset of the original training set.

6. Return to Step 2, above. Add the new PRESS value calculated in step 3, to the PRESS values calculated so far. Continue this process until PRESS values for all combinations of "leave one out" have been computed and summed.

Steps 1 - 6 are repeated for calibrations generated with every possible rank (number of factors). We can then examine the PRESS for each of the calibrations and choose the one that gives the best results. The number of factors used in that calibration is the rank of the system.

This procedure is known as "leave one out" cross-validation. This is not the only way to do cross-validation. We could apply this approach by leaving out all permutations of any number of samples from the training set. The only constraint is the size of the training set, itself. Nonetheless, whenever the term cross-validation is used, it almost always refers to "leave one out" cross-validation.

Regenerate the Data

As we saw in the last chapter, by discarding the noise eigenvectors, we are able to remove a portion of the noise from our data. We have called the data that results after the noise removal the *regenerated data*. When we perform principal component regression, there is not really a separate, explicit data regeneration step. By operating with the new coordinate system, we are automatically regenerating the data without the noise.

Computing the calibration

We compute a PCR calibration in exactly the same way we computed an ILS calibration. The only difference is the data we start with. Instead of directly using absorbance values expressed in the spectral coordinate system, we use the same absorbance values but express them in the coordinate system defined by the basis vectors we have retained. Instead of a data matrix containing absorbance values, we have a data matrix containing the coordinates of each spectrum on each of the axes of our new coordinate system. We have seen that these new coordinates are nothing more than the projections of the spectra onto the basis vectors. These projections are easily computed:

$$\mathbf{A_{proj}} = \mathbf{Vc^T}\,\mathbf{A} \qquad [58]$$

where:

$\mathbf{A_{proj}}$ is the matrix containing the new coordinates (the projections)

A   is the original training set absorbance matrix

Vc   is the matrix containing the basis vectors, one column for each
        factor retained.

Now we can substitute A_{proj} into equation [46] in place of A:

$$C = F A_{proj} \qquad [59]$$

We have also changed the name of the regression matrix to **F** in order to distinguish it from the ILS regression matrix, **P**, in equation [46].

Now we are ready to solve for the PCR calibration matrix. We do this exactly the same way we solved for the ILS calibration. First, we post-multiply both sides of equation [59] by A^{T}_{proj}.

$$C A^{T}_{proj} = F A_{proj} A^{T}_{proj} \qquad [60]$$

Next, we post-multiply both sides of equation [60] by $[A_{proj} A^{T}_{proj}]^{-1}$, the pseudo-inverse of A^{T}_{proj}.

$$C A^{T}_{proj} [A_{proj} A^{T}_{proj}]^{-1} = F A_{proj} A^{T}_{proj} [A_{proj} A^{T}_{proj}]^{-1} \qquad [61]$$

Since the product of a matrix and its inverse is the identity matrix, the quantity $A_{proj} A^{T}_{proj} [A_{proj} A^{T}_{proj}]^{-1}$ disappears from the right-hand side of equation [61], leaving:

$$C A^{T}_{proj} [A_{proj} A^{T}_{proj}]^{-1} = F \qquad [62]$$

Predicting Unknowns

Now that we have calculated **F**, we can use it to predict the concentrations in an unknown sample from its measured spectrum. First, we substitute the expression for A_{proj} from equation [58] into equation [59], adding subscripts to indicate we are predicting the concentrations for an unknown sample:

$$C_{unk} = F Vc^{T} A_{unk} \qquad [63]$$

Notice that we can pre-calculate the quantity $\mathbf{Vc^T\,A}$ at calibration time. Let's call the result $\mathbf{F_{cal}}$. Equation [63] becomes

$$\mathbf{C_{unk}} = \mathbf{F_{cal}\,A_{unk}} \qquad\qquad [64]$$

The calibration matrix, $\mathbf{F_{cal}}$ has exactly the same format as $\mathbf{K_{cal}}$, the calibration matrix for CLS. It has one row for each component being predicted. Each row has one calibration coefficient for each wavelength in the spectrum. We can now use $\mathbf{F_{cal}}$ to predict the concentrations in an unknown sample from its measured spectrum. First, we place the spectrum into a new absorbance matrix, $\mathbf{A_{unk}}$. We can now use equation [64] to produce a new concentration matrix, $\mathbf{C_{unk}}$, containing the predicted concentration values for the unknown sample.

That's all there is to it. In the next chapter, we'll see how it works on our simulated data.

PCR in Action

Now, we are ready to apply PCR to our simulated data set. For each training set absorbance matrix, **A1** and **A2**, we will find all of the possible eigenvectors. Then, we will decide how many to keep as our basis set. Next, we will construct calibrations by using ILS in the new coordinate system defined by the basis set. Finally, we will use the calibrations to predict the concentrations for our validation sets.

All we need do to calculate all the possible eigenvectors and eigenvalues is feed the data into an appropriate software package. So, we will begin the discussion with the question of how many of the eigenvectors to keep.

Choosing the Optimum Rank

Table 8 contains the eigenvalues (EV's) and reduced eigenvalues (REV's) that we get for the data in our training set absorbance matrices, **A1**, and **A2**. These are also plotted in Figure 53.

When we look at Table 8 and Figure 53 it is apparent that something changes when moving from the 5th eigenvalue to the 6th. At that point hin the

	A1		A2	
Rank	EV	REV	EV	REV
1	586.0547	0.3907	451.8087	0.3012
2	27.3511	0.0197	18.1739	0.0131
3	9.2579	0.0073	7.0551	0.0055
4	2.5079	0.0022	1.5713	0.0013
5	1.0125	0.0010	1.2394	0.0012
6	0.3308	0.0003	0.4050	0.0004
7	0.3208	0.0004	0.3404	0.0004
8	0.2864	0.0004	0.3213	0.0004
9	0.2609	0.0004	0.2901	0.0005
10	0.2429	0.0004	0.2401	0.0004
11	0.2212	0.0005	0.2263	0.0005
12	0.2114	0.0006	0.1939	0.0005
13	0.1835	0.0007	0.1647	0.0006

Table 8. Eigenvalues (EV) and reduced eigenvalues (REV) for the two training sets, **A1** and **A2**.

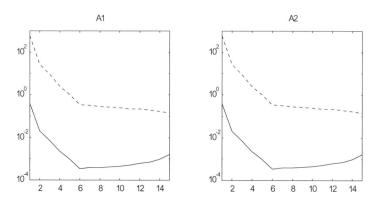

Figure 53. Logarithmic plots of eigenvalues (- -) and reduced eigenvalues (---) for the two training sets, **A1** and **A2**.

sequence, the rate that the eigenvalues decrease with increasing rank suddenly becomes smaller. The inflection point is even more apparent when we look at the REV's. The REV for rank 7 is essentially the same as the REV for rank 6. In other words, eigenvectors 1 - 5 appear to have captured a larger equivalent share of variance than eigenvectors 6 and higher. Based on this "eyeball" inspection of the EV's and REV's, we would estimate that our data has an intrinsic dimensionality, or rank, or 5. Stated another way, there appear to be 5 independent underlying sources contributing to the systematic (non-noise) variations in our data.

Let's compare these plots of the REV's to the plot in Figure 52. Notice that these REV's do not exhibit ideal behavior. Ideally, as rank increases, the REV's would drop to some minimum value and then remain at that level. These REV's begin to tail back up. This sort of non-ideal behavior is not uncommon when working with actual data. Unfortunately, it can complicate matters when we use the 2-way F-test to see which REV's represent basis vectors and which ones represent noise vectors.

We have an advantage in this situation. We know that our data contain 4 components, plus a small linear baseline, and some nonlinearities. This would lead us to expect that our data would show a rank of 5 or 6. Four dimensions are required to span the variations from the 4 different spectral components, and an additional 1 or two would be needed to span the variations due to the baselines and non-linearities. It gets a bit tricky when we realize that the baseline and non-linearity effects are rather small. Statistical indicators such as the 2-way F-test on REV's are not always able to destinguish factors that span small, but systematic, variations in the data from factors that span pure random noise. For this reason, PRESS of validation data or cross-validation is sometimes a more reliable way to make the determination.

Let's see what the 2-way F-test tells us. Table 9 contains the REV's and F ratios for the data in the two training sets **A1** and **A2**. The details of calculating the F ratios, and determining the values for the numerator and denominator are discussed in Appendix D.

Remember, this test is asking the question, "Is the reduced eigenvalue, n, (statistically) significantly greater than the reduced eigenvalue, n + 1?" We begin asking the question at the bottom of the table. We see that the F values for **A1** and **A2** at rank 14 are less than the values from the statistical tables. Thus, the answer for rank 14 is "no." We continue upwards until we find the rank for which the F exceeds the value from the statistical table at the desired level of significance. At that point, we keep that eigenvector and all the ones that lie higher on the table, and we stop the test. Once we reach a significant eigenvalue, the F values for all of the reduced eigenvalues situated above it on the table are not valid for purposes of the test. When we examine the values of F

Rank	Numerator	Denominator	A1 REV	A1 F	A2 REV	A2 F	F at 5%	F at 10%
1	1	14	0.3907	91.3900	0.3012	98.3185	4.60	3.10
2	1	13	0.0197	11.1469	0.0131	9.1354	4.67	3.14
3	1	12	0.0073	8.9883	0.0055	7.7190	4.75	3.18
4	1	11	0.0022	3.9011	0.0013	2.2609	4.84	3.23
5	1	10	0.0010	2.0512	0.0012	2.4620	4.96	3.29
6	1	9	0.0003	0.7035	0.0004	0.8729	5.12	3.36
7	1	8	0.0004	0.7221	0.0004	0.7879	5.32	3.46
8	1	7	0.0004	0.6797	0.0004	0.8089	5.59	3.59
9	1	6	0.0004	0.6515	0.0005	0.8009	5.99	3.78
10	1	5	0.0004	0.6406	0.0004	0.7173	6.61	4.06
11	1	4	0.0005	0.6158	0.0005	0.7512	7.71	4.54
12	1	3	0.0006	0.6336	0.0005	0.7221	10.13	5.54
13	1	2	0.0007	0.5875	0.0006	0.7039	18.51	8.53
14	1	1	0.0010	0.6067	0.0008	0.8612	161.4	39.86

Table 9. REV's and F for the two training sets **A1** and **A2**. F values at 5% and 10% are from standard statistical tables.

for **A1,** we see that the value at rank 3 exceeds the F value from the statistical tables for the 5% level. This means that there is less than a 5% probability that the 3rd reduced eigenvalue belongs to a noise eigenvector. We also see that the F value at rank 4 exceeds the F value from the statistical tables for the 10% level. This indicates that there is less than a 10% probability that the 4th reduced eigenvalue belongs to a noise eigenvector. If we work with the F values for **A2,** at either significance level, we are led to discard all but the first 3 eigenvectors. These results might seem surprising, considering what we know about the data. The problem is that the F ratios are being skewed by the nonideal behavior of the reduced eigenvalues at the bottom of the table. We can show this by artificially setting the REV's for rank 10 and higher to equal .0004, and recalculating the F ratios. The resulting F values for the first 7 REV's are shown in Table 10.

At the 10% level, these modified F ratios indicate that we need 5 factors for either **A1** or **A2.** At the 5% level, we get an indication of 4 factors for **A1** and 3 factors for **A2.**

This lack of sharpness of the 1-way F-test on REV's is sometimes seen when there is information spanned by some eigenvectors that is at or below the level of the noise spanned by those eigenvectors. Our data sets are a good example of such data. Here we have a 4 component system that contains some nonlinearities. This means that, to span the information in our data, we should expect to need at least 4 eigenvectors — one for each of the components, plus at least one additional eigenvector to span the additional variance in the data caused by the non-linearity. But the F-test on the reduced eigenvalues only

Rank	Numerator	Denominator	A1 REV	modified A1 F	A2 REV	modified A2 F	F at 5%	F at 10%
1	1	14	0.3907	93.8416	0.3012	101.6536	4.60	3.10
2	1	13	0.0197	12.0286	0.0131	9.9446	4.67	3.14
3	1	12	0.0073	11.0777	0.0055	9.5448	4.75	3.18
4	1	11	0.0022	5.8109	0.0013	3.1125	4.84	3.23
5	1	10	0.0010	3.8654	0.0012	4.2027	4.96	3.29
6	1	9	0.0003	1.5485	0.0004	1.7379	5.12	3.36

Table 10. First 7 REV's and recalculated F for the two training sets **A1** and **A2.** F values at 5% and 10% are from standard statistical tables.

considers the magnitude of the variance spanned by each of the eigenvectors. Thus, if we have a nonlinearity that introduces additional variance that is at or below the noise level, the F-test may not provide the best guidance on the number of factors to keep.

Fortunately, since we also have concentration values for our samples, We have another way of deciding how many factors to keep. We can create calibrations with different numbers of basis vectors and evaluate which of these calibrations provides the best predictions of the concentrations in independent unknown samples. Recall that we do this by examing the Predicted Residual Error Sum-of Squares (PRESS) for the predicted concentrations of validation samples.

Figure 54 contains plots of the PRESS values we get when we use the calibrations generated with training sets **A1** and **A2** to predict the concentrations in the validation set **A3**. We plot PRESS as a function of the rank (number of factors) used for the calibration. Using our system of nomenclature, the PRESS values obtained by using the calibrations from **A1** to predict **A3** are named PCRPRESS13. The PRESS values obtained by using the calibrations from **A2** to predict the concentrations in **A3** are named PCRPRESS23. It is clear from the plots that the errors in predicting the concentrations of the validation set, **A3**, are minimized by using 5 factors for the calibration.

If we did not have a validation set available to us, we could use cross-validation for the same purposes. Figure 55 contains plots of the results of cross validation of the two training sets, **A1** and **A2**. Since no separate validation data set is involved, we name the results PCRCROSS1 and PCRCROSS2, respectively.

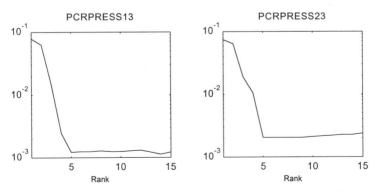

Figure 54. Logarithmic plots of the PRESS values as a function of the number of factors (rank) used to construct the calibration.

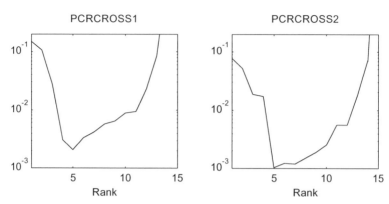

Figure 55. Logarithmic plots of the cross-validation results as a function of the number of factors (rank) used to construct the calibration.

Again, it is clear from these plots that the errors are minimized when 5 factors are used. Thus, we will construct our calibration matrices using a basis space cromprised of the first 5 eigenvectors (factors).

The plots in Figure 56 complete the story. They show why, if we do not have validation samples available, we cannot simply use the fits to the training set to determine how many factors to keep. First let's recall that, by fits to the training set, we mean the procedure where we generate a calibration from a training set and use that calibration to predict the concentrations of the samples in that same training set. We then examine the PRESS for these predictions for an indication of how well the calibration was able to fit the data in the training set. Figure 56 contains plots of the fits for the two training sets, **A1** and **A2**. Since there are no independent validation sets are involved, we have named the results PCRPRESS1 and PCRPRESS2, respectively.

When we examine the plots in Figure 56 we see that the PRESS decreases each time we add another factor to the basis space. When all of the factors are included, the PRESS drops all the way to zero. Thus, these fits cannot provide us with any information about the dimensionality of the data. The problem is that we are trying to use the same data for both the training and validation data. We lose the ability to assess the optimum rank for the basis space because we do not have independent validation samples that contain independent noise. So, the more factors we add, the better the calibration is able to model *the particular noise* in these samples. When we use all of the factors, we are able to model the noise *completely*. Thus, when we predict the concentrations for

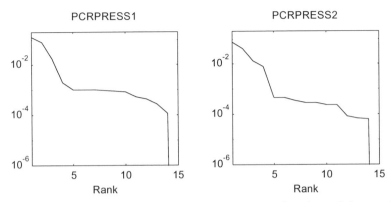

Figure 56. Logarithmic plots of the PRESS values as a function of the number of factors (rank) using the same samples for calibration and validation. As factors are added, the errors continue to decrease. When all of the factors are used, the errors equal exactly zero.

these exact same samples that contain exactly the same noise, our PRESS values decrease each time a factor is added. When all of the factors are used, the noise is modeled completely, and the PRESS values drop to zero. So if we are trying to find the correct number of factors to use, and we do not have independent validation samples, we must use a technique such as cross-validation. Simple fits to the training set are useless for this purpose.

Basis Vectors and Noise Vectors

So, cross-validation and PRESS both indicate that we should use 5 factors for our calibrations. This indication is sufficiently consistent with the F-test on the REV's and with our "eyeball" inspection of the EV's and REV's, themselves. It can also be worthwhile to look at the eigenvectors themselves.

The eigenvectors are called *abstract factors* because these axes of our new, optimum, coordinate system for the data are chosen without regard for their physical or chemical significance. These axes were selected because they are the most efficient way of spanning the variance in the data, and not because they are aligned with the pure component spectra of the components or with some other meaningful parameter of the data. (There is an entire field called *Target Factor Analysis* which concerns itself with transforming the abstract factors into physically or chemically meaningful factors.) Nonetheless, each factor is nothing more than some unique axis in the original absorbance space. (Recall that the factors are all normalized to unit length.) Referring back to

Figure 45, we see that the end point of each factor can be expressed in terms of its coordinates along the original absorbance axes. If our spectra were measured at 100 wavelengths, the coordinates of the end point of each factor could be expressed as 100 absorbances values, one for each of the original wavelengths. Thus we can plot each factor just as if it were a spectrum. But since these are abstract spectra, we should not expect that they will, generally, look like recognizable spectra. Figure 57 contains plots of the first 4 factors of the two training sets **A1**, and **A2**. We have named the matrices holding these factors **Vc1** and **Vc2**, respectively.

In Figure 57, we notice that the first factors for each training set are quite similar to each other. Also, they do not look all that abstract. Since we did not scale or center the spectra in the training sets prior to analysis, the first factor for each absorbance matrix is the least-squares average of all the spectra in the set. The second factors, on the other hand, appear a bit more abstract, although they contain spectral like features. In fact, they look a bit like first derivative spectra. This is to be expected since the second factors are capturing the maximum amount of the residual variance left behind by the first factors. This residual variance is basically the least-squares average of the differences between all of the spectra and the first factor in each data set. So it should be no surprise that the second factors have these kinds of features. It should also be expected that these factors will begin to look different from each other, because the factors are now starting to pick up the individual differences between the spectra in **A1** and those in **A2**.

Moving down to the third factors, we still see spectral like features, but now the differences between the factors for each data set are becoming more noticable. Also, note that these two factors are roughly mirror images of each other with respect a line through 0 absorbance. When we recall that the sign of the vectors is arbitrary, we realize that this is nothing to worry about. Sometimes changing a single sample in a training set will cause the sign of a factor to change from positive to negative.

The fourth factors are much more different from each other than were the first three. They still contain strong, spectral-like features, but they look a bit noisier that the earlier factors. So far, all of these factors look like they could be spanning significant information. Let's look at a few more factors which we have plotted in Figure 58.

In Figure 58 the fifth factors appear quite noisy. Nonetheless, we can imagine that there are still some systematic features in these factors. The fact that these apparent features are not much stronger than the noise is consistent with the results of the F-tests on the REV's. It can be dangerous to decide

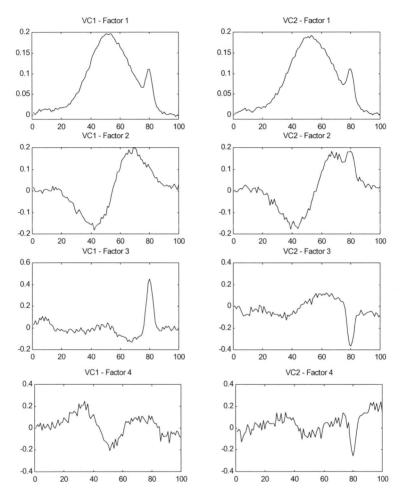

Figure 57. First 4 factors for the two training sets, **A1** and **A2**.

whether or not to retain a factor by inspecting it visually. It is too easy to see patterns in data that are, in fact, random. (For example, when we look at the sky we can see constellations). But, we based our decision to retain this factor on the fact that we got the lowest PRESS and cross-validation values with 5 factors. The fact that we can see features on this factor serves to increase our confidence in the decision.

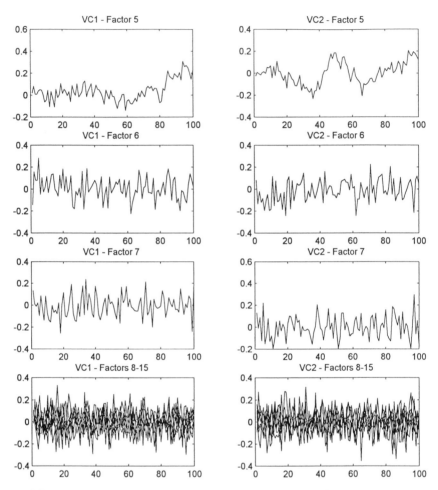

Figure 58. Remaining 11 factors for the two training sets, **A1** and **A2**.

All of the remaining factors do appear to contain nothing but noise. Remember that true noise eigenvectors will lie in some random direction that is devoid of any useful information. Thus, they should look like pure noise.

Just for fun, let's look at the distribution of the absorbances in each factor. Figure 59, contains histograms of the absorbances in the first 8 factors for the first training set. If a factor is purely a noise factor, it's absorbances should follow a gaussian distribution. The absorbances of the first 4 factors do appear to deviate significantly from a gaussian distribution. Notice that, since our data

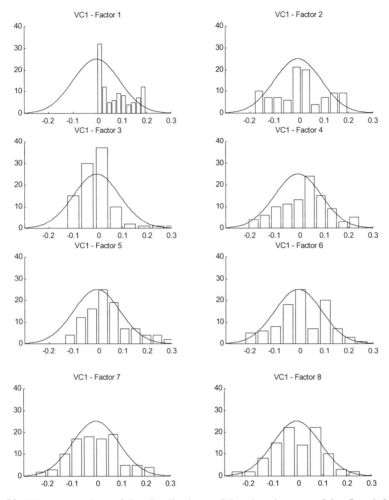

Figure 59. Histogram plots of the distributions of the absorbances of the first 8 factors of the training set, **A1**. A plot of an ideal gaussian distribution is superimposed on each histogram.

was not mean centered, the first factor removed the mean from the data. Thus, all the subsequent factors are mean centered. The absorbances of factors 8, 7, and even 6 appear to be reasonably gaussian in their distribution. And, with the benefit of hindsight, we are very tempted to conclude that the distribution of the absorbances of factor 5 deviate significantly from gaussian. Naturally, visual inspection is not a sound way to draw any conclusions from these plots.

However, a discussion of an analytical approach to this issue is beyond the scope of this text.

Regenerated Data and Residuals

We've seen that the data regeneration step is implicit in the calibration. Even though there is no need to explicitly regenerate the data, it is, nonetheless instructive. Let's use the 5 basis vectors for training set **A1** to regenerate the spectra in **A1**. Let's also look at the residuals, that portion of the variance that is discarded from the regenerated data because it is (hopefully) pure noise. We will name the matrix holding the regenerated spectra and the residuals PCAREG1 and PCARESID1, respectively. Figure 60 contains a plot of one of these regenerated spectra together with a plot of the same spectrum before regeneration. We can easily see that a significant amount of noise has been removed without any evident degradation of the spectrum.

Figure 60 also contains a plot of the differences between the original and the regenerated spectra. This is identical to the residuals. The residuals of

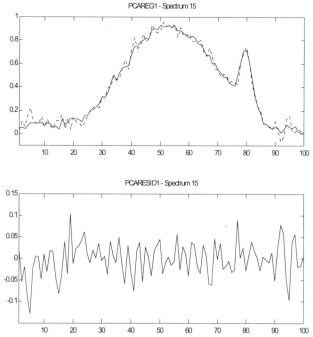

Figure 60. Plot (top) of a regenerated spectrum (——) and the original spectrum (---) of a sample in training set A1 together with (bottom) a separate plot of the differences between the two spectra.

this spectrum look comfortably like pure random noise. Figure 61 contains plots of the residuals for all of the spectra in **A1** and **A2**. Here too, the residuals do not appear to have any spectral-like features.

We can also use the 5 factors which comprise the basis space of **A1** to regenerate the spectra in our three validation sets **A3**, **A4**, and **A5**. We will name the matrices holding these spectra, PCAREG13, PCAREG14, PCAREG15 and PCARESID13, PCARESID14 and PCARESID15, respectively.

Figure 62 contains plots of one regenerated spectrum from each validation set together with the same spectrum before regeneration. Figure 62 also contains plots of the residuals of all of the regenerated spectra in these validation sets. In Figure 62 , we can see that the basis space of our training set does a fine job of regenerating the validation spectra in **A3**. Noise is nicely

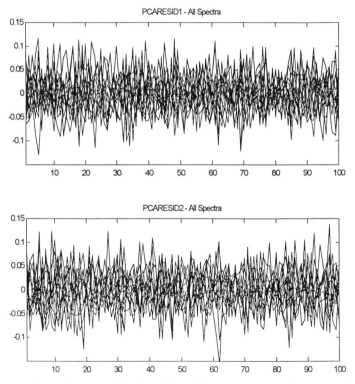

Figure 61. Plot of the residuals of the all the regenerated spectra in the two training sets, **A1** and **A2**.

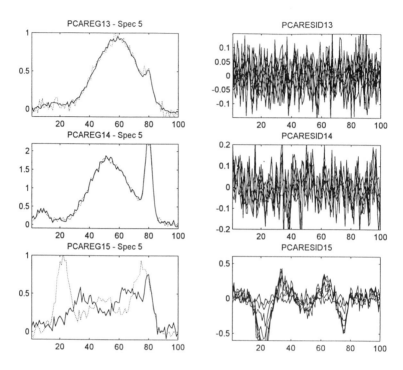

Figure 62. Plots (left column) of a regenerated spectrum (—) and the original spectrum (···) of a sample from each of the 3 validation sets **A3**, **A4**, and **A5**. Residuals for all of the regenerated spectra in each of the 3 validation sets are also shown (right column).

removed without any noticable degradation of the spectrum. Notice that the residuals for **A3** do not seem to contain any spectral features, and they are roughly the same magnitude as the residuals for **A1**.

Regeneration also seems to work well on the validation spectra in **A4**. Again, we see good noise removal without spectral degradation. Notice, however, that the residuals for **A4** are a bit larger in magnitude than those of **A1**. Also, there appears to be a small amount of spectral-like structure in the residuals. This makes sense when we remember that the samples in **A4** are the overrange samples. Thus, they are somewhat different from the samples in the training set that was used to develop the basis space. So, it makes sense that the residuals are a bit higher. The spectral-like features are present in the residuals because the nonlinearity affects the samples in **A4** more than in the training set.

So, it is not unusual if basis vectors calculated for **A1** do not completely span all of the variance which is due to the increased nonlinearities in the **A4** samples.

The story is very different for the samples in **A5**. Here, we see that the regenerated spectrum has major differences from the original spectrum. Also, the residuals are much larger and display significant amounts of spectral-like features. Of course, the reasons for this are simple. **A5** contains samples with varying amounts of an additional, interfering component. The samples in our training set, **A1**, do not contain any of this interfering component. So there is no way that the basis vectors for the **A1** spectra can span *all* of the variance added to the **A5** spectra by that component. So it makes sense that major spectral features are missing from the regenerated spectra and show up, instead, in the residuals.

Confidence Indicator

So, we see that the spectral residuals for a sample will be higher whenever there is something that introduces a mode of variation into the spectrum that was not present in any of the training samples used to develop the basis space. The anomolous variation could be caused by instrument drift, an unexpected interfering component, a misaligned sample cell, or whatever. We can use this property of residuals as an indicator that can signal us whenever a sample is significantly different from the training set samples. This is very valuable because if we try to predict the concentrations of a sample that differs significantly from the samples with which the calibration was generated, the reliability of the predictions is very poor.

We can use the sum of the squares of the residuals (SSR) of the training set as our benchmark. Then, we can establish one or more confidence limits based on this benchmark. Typically, we might set a warning level at 2 to 3 times the training set SSR. Anytime the residuals of an unknown spectrum exceed the warning level we could take appropriate action. We might turn on a yellow light, issue a warning message, send an e-mail to the person responsible for the analysis, repeat the measurement, capture a sample, save a spectrum to disk, initiate a self-diagnostic routine for the analyzer, or whatever. We could also set an alarm level. Typically this would be set at 3 to 4 times the training set SSR. If the SSR of an unknown exceeded the alarm level we could turn on a red light, sound an alarm, save the data to disk, capture a sample, initiate self-diagnostics, refuse to report the predicted concentration values, or shut down the analyzer. Table 11 shows the SSR's for training set **A1**, and the three validation sets, **A3** through **A5**.

Data Set	A1	A3	A4	A5
SSR	0.1577	0.3117	0.6364	3.7120

Table 11. Sum of the square of residuals (SSR) for **A1** and **A3** through **A5**, using the 5 basis vectors for **A1**.

Suppose we set our warning and alarm levels at typical levels of 3 and 5 times the training set SSR, respectively. We can see in Table 11 that our green light would stay on while predicting the samples in the normal validation set, **A3**. If we encounter samples from **A4**, the overrange validation set, the yellow light would come on. And when we see samples from **A5**, the validation set with the unexpected interfering component, red lights should flash, alarms should sound, etc.

Table 12 shows the SSR's for each sample in validation set **A5** together with the concentration of the unexpected component in each sample. Figure 63 contains a plot of the data in Table 12.

We can see, in Table 12 that there is a monotonic relationship between the SSR and the concentration of the interferring 5th component in each sample of the validation set **A5**. In Figure 63 we can see that the relationship is approximately linear with the square root of the SSR. The important thing is not the linearity of the relationship, but that it exists at all and increases monotonically. It gives us a very useful way of flagging samples which our

Data Set	A1 SSR	A5 Conc.	A5 SSR
	0.1577	0.9880	7.6866
		0.9353	6.7935
		0.8144	5.4329
		0.7733	4.7334
		0.6074	3.0793
		0.3177	1.1992
		0.1161	0.4636

Table 12. Sum of the square of residuals (SSR) for the individual samples in the validation set, **A5**, using the 5 basis vectors for **A1** together with the concentrations of the unexpected 5th component in the **A5** samples.

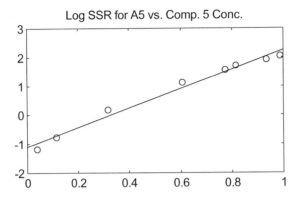

Figure 63. Semi-logarithmic plot of the SSR (y-axis) vs. the concentration of Component 5 (x-axis) for each sample in A5.

calibration may not be able to handle properly. This capability, alone, will usually give us sufficient reason to use the factor-based techniques to develop our calibrations. As we are about to see, calibrations produced with the factor-based techniques also tend to perform better than ordinary CLS or ILS calibrations.

PCR Calibration Matrices

First, let's look at the PCR regression coefficients in the calibration matrices we produce from the two training sets, **A1** and **A2**. We will name these calibration matrices $F1_{cal}$ and $F2_{cal}$, respectively. Recall that the calibration matrices have a row for each component being predicted. Each row has one regression coefficient for each spectral wavelength. Thus, we can plot each row of the regression matrix as if it were a spectrum. Figure 64 contains these plots. We can think of these plots as the "strategy" of the calibration. They show us which wavelengths are taken as positively correlated with the predicted concentrations, which negatively, and which wavelengths are essentially ignored.

One of the first things we notice is that the regression coefficients for each component produced by the two training sets are quite similar to each other. Contrast this with the CLS calibration coefficients plotted in Figure 21. It is also apparent that the coefficients are reasonably well conditioned. In other words, their magnitude is not excessive, and they do not swing wildly from large positive to large negative values.

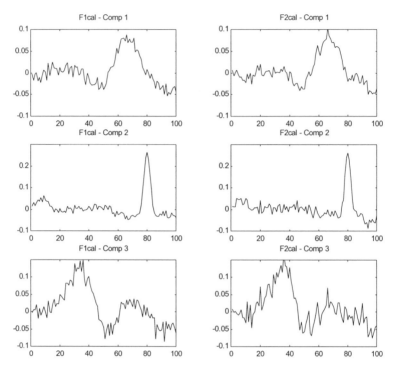

Figure 64. Plots of the PCR calibration coefficients calculated for each component with each training set.

PCR Predictions on the Validation Sets

Let's see how well the PCR calibrations predict the concentrations of our 3 validation sets **A3** - **A5**. We do this by substituting **A3** - **A5** into equation [64], first with $\mathbf{F1}_{cal}$, then with $\mathbf{F2}_{cal}$ to produce 6 concentration matrices containing the estimated concentrations. We will name these matrices $\mathbf{F13}_{res}$ through $\mathbf{F15}_{res}$ and $\mathbf{F23}_{res}$ through $\mathbf{F25}_{res}$. Using this naming system, $\mathbf{F24}_{res}$ is a concentration matrix holding the concentrations for validation set **A4** predicted with the calibration matrix $\mathbf{F2}_{cal}$, that was generated with training set **A2**, the one which was constructed with the random design. Again, there is data "crib sheet" inside the back cover to help you keep things straight. Figure 65 contains plots of the expected vs. predicted concentrations for $\mathbf{F13}_{res}$ through $\mathbf{F25}_{res}$. Table 13 contains the values for PRESS, SEC^2, SEP^2, and r, for this set of results.

	$F1_{cal}$			$F2_{cal}$		
	PRESS	SEC²	r	PRESS	SEC²	r
A1	.0010	.0016	.9970	-	-	-
A2	-	-	-	.0005	.0008	.9974
A3	.0012	.0013	.9943	.0021	.0022	.9898
A4	.0034	.0032	.9987	.0063	.0070	.9961
A5	.1277	.1367	.8819	.0914	.1020	.9133

Table 13. PRESS, SEC², SEP², and r for $F1_{res}$ through $F25_{res}$.

It is apparent that these are the best prediction results for this data that we have seen up to this point. We even do extremely well with the overrange validation samples in **A4**. But, it would be dangerous to assume that we can routinely get away with extrapolation of this kind. Sometimes it can be done, sometimes it can't. There is no simple rule that can tell us which situation we might be facing. It is very dependent on the particular data and application involved. In any case, it is usually a good policy to strongly discourage using a calibration to predict the concentrations in samples that require extrapolation.

Of course, the calibrations do rather poorly predicting the concentrations of the samples in **A5**. This is exactly as expected since these samples have varying amounts of an additional, unexpected component that wasn't present in any of the calibration samples. But, with the factor-based techniques, we have the ability to detect these samples using the SSR's of the spectra. So if we encounter any unknowns for which the calibration must be considered invalid, we now know how to take appropriate action.

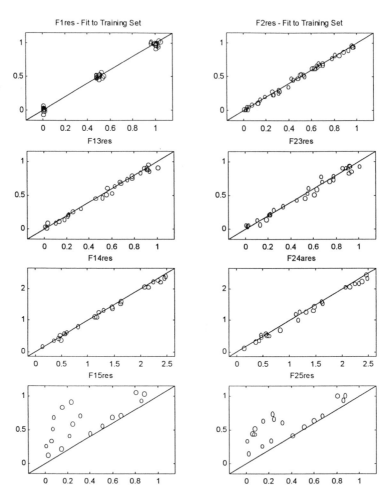

Figure 65. Expected concentrations (x-axis) vs. predicted concentrations (y-axis) for PCR calibrations (see text).

Partial Least-Squares

Partial least-squares in latent variables (PLS) is sometimes called partial least-squares regression, or PLSR. As we are about to see, PLS is a logical, easy to understand, variation of PCR.

We have seen that PCR is simply ILS performed using a different, optimally efficient coordinate system for the spectra. PLS takes this concept one step further by using a different strategy to find a coordinate system that can have advantages over the coordinate system used for PCR. This strategy involves finding factors for both the spectral *and* the concentration data. The reasoning behind this approach is twofold. First, why not utilize the noise removal capabilities of PCA to remove some of the noise from the concentration data? Second, we have seen how the errors due to the noise in the spectral data will, in general, deflect each eigenvector slightly out of the plane containing the theoretical, noise-free data. Since the noise in the concentration data is independent of the noise in the spectral data, the errors due to the noise in the concentration data will, in general, deflect each concentration eigenvector in some randomly different direction than the deflection of the corresponding spectral eigenvector. So, if we compare each spectral vector with its corresponding concentration vector, they will have some (hopefully small) angle between them. Since this angle is due to the differences in the particular noise between the two data spaces, why not rotate the vectors back toward each other until they are aligned? In general, this rotation should provide better noise removal by bring the vectors closer to the ideal planes containing the noise-free spectral and concentration data.

PLS vs. PCR: Similarities and Differences

Let's summarize the similarities of and differences between PCR and PLS. When we do PLS, we follow exactly the same steps for PCR, as shown in Figure 49, except:

1. The data matrices generally use the row-wise convention.
2. In addition to the set of new coordinate axes (basis space) for the spectral data (the x-block), we also find a set of new coordinate axes (basis space) for the concentration data (the y-block).
3. In addition to expressing the spectral data as projections onto the spectral factors (basis vectors), we express the concentration data as projections onto the concentration factors (basis vectors).

4. On a rank-by-rank (i.e. factor-by-factor) basis, we rotate, or perturb, each pair of factors, (1 spectral factor and its corresponding concentration factor) towards each other to maximize the fit of the linear regression between the projections of the spectra onto the spectral factor with the projections of the concentrations onto the concentration factor.

5. We calculate the calibration (regression) coefficients on a rank-by-rank basis using linear regression between the projections of the spectra on each individual spectral factor with the projections of the concentrations on each corresponding concentration factor of the same rank.

The prediction step for PLS is also slightly different than for PCR. It is also done on a rank-by-rank basis using pairs of spectal and concentration factors. For each component, the projection of the unknown spectrum onto the first spectral factor is scaled by a response coefficient to become a corresponding projection on the first concentration factor. This yields the contribution to the total concentration for that component that is captured by the first pair of spectral and concentration factors. We then repeat the process for the second pair of factors, adding its concentration contribution to the contribution from the first pair of factors. We continue summing the contributions from each successive factor pair until all of the factors in the basis space have been used.

Visualizing PLS

If you have not yet read the chapter on Factors Spaces, or if your recollection of that chapter is at all hazy, you would probably find it useful to review that chapter before proceeding beyond this point. We are going to use a similiar graphical approach to understand how PLS works.

PCA of Both the Spectral and the Concentration Data

We've said that PLS involves finding a set of basis vectors for the spectral data *and* a separate set of basis vectors for the concentration data. So, we need to understand how the spectral factors and the concentration factors are related to each other.

Let's consider the same set of perfectly linear, noise-free data that was introduced in Figure 32. There is no need to start flipping pages; we'll reproduce Figure 32 here as Figure 66. Figure 67 contains two additional views of the data in Figure 67. In Figure 68 we will also plot the compostitions of the samples whose spectra are plotted in Figures 66 and 67. The concentrations of the first component are plotted along one axis, and the concentrations of the second component are plotted along the other axis.

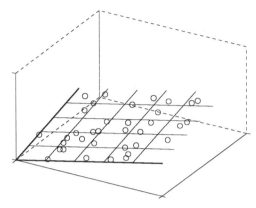

Figure 66. Multivariate plot of 3-wavelength spectra for samples containing varying amounts of Component 1 and Component 2. In this noise-free, linear case, all of the specra must lie in a plane.

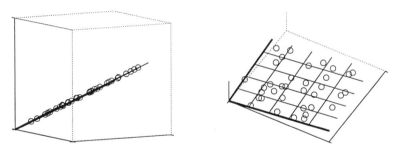

Figure 67. Two different views of the data in Figure 66.

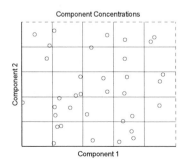

Figure 68. Plot of the component concentrations for the samples in Figure 66.

Recall that Figures 66 and 67 contain plots of the spectra of a number of samples containing varying amounts of 2 different components. The spectra are measured at 3 wavelengths. For each spectrum, the absorbance at the first wavelength is plotted along one axis, the absorbance at the second wavelength is plotted along another axis, and the absorbance at the third wavelength is plotted along the remaining axis. If you have any questions about why all of these noise-free, perfectly linear spectra must lie exactly in a plane, and why that plane is oriented at some angle in the 3-dimensional absorbance space, then please don't try to read any further. Go back and study the chapter on Factor Spaces until you understand these issues before continuing beyond this point.

Figure 68 contains a plot of the composition of the samples whose spectra are plotted in Figure 67. For each sample, the concentration of the first component is plotted along one axis and the composition of the second component is plotted along the second axis. Examining Figures 66 through 68, it is immediately apparent that the relative positions of the points in Figure 68 are identical to the relative positions of the points within the plane in Figures 66 and 67. In other words, if we were to appropriately scale each concentration axes in Figure 68, we could take the plot from Figure 68 and lay it onto the plane containing the data points in Figures 66 and 67 in such a way that the points in both plots will lie exactly on top of one another. To show this clearly, we rotate Figure 67 so the plane containing the spectral data points is flat on the page, and plot it in Figure 69 side-by-side with the concentration data. Stated yet another way, the points in Figure 68 are congruent with the points in Figures 66 and 67. The points must be congruent because the concentrations plotted in Figure 68 were used to create the spectra in Figure 66 with perfect linearity, and without any random noise. If the truth of this is not obvious to you, then please review the chapter on Factor Spaces.

Now, let's calculate the eigenvectors for the spectra in Figure 66. In fact, we've already done this in the chapter of Factor Spaces. They were plotted in Figures 37 and 38. For convenience, we reproduce these plots here as Figures 70 and 71.

In order to process these data with PLS, we must also calculate the eigenvectors for the concentration data in Figure 68. Figure 72 contains plots of the first two eigenvectors for the concentration data overlaid onto the plot of the concentration data points themselves. Recall that eigenvectors are always normalized to unit length. However, we have plotted the eigenvectors with different lengths for clarity.

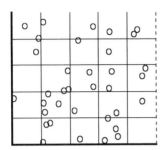

Figure 69. Spectral and concentration data plotted side-by-side to show the congruence of the points in the two different data spaces.

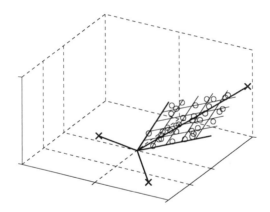

Figure 70. The spectral data from Figure 66 plotted together with all 3 eigenvectors (factors) for the data.

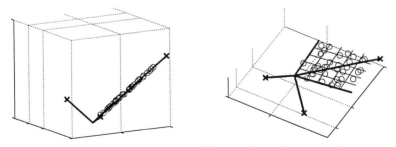

Figure 71. The spectral data from Figure 66 plotted together with all 3 eigenvectors (factors) for the data.

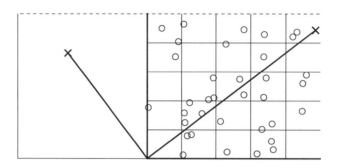

Figure 72. The concentration data from Figure 68 plotted together with the first 2 eigenvectors (factors) for the data. The eigenvectors are shown as having different lengths for clarity. In reality they both have unit length.

It is evident in Figures 70 through 72 that the first two eigenvectors of the concentration data are congruent with the first two eigenvectors of the spectral data. Notice that the second concentration vector and the second spectral vector happen to have opposite signs. But, recall that this is of no consequence because the sign of an eigenvector is completely arbitrary.

Just as the spectral and concentration data points are exactly congruent with each other within the planes containing the data points, the spectral and concentration eigenvectors for this noise-free, perfectly linear case must also be exactly congruent. Because the vectors are congruent, the projection of each spectral data point onto a spectral factor must be directly proportional to the projection of the corresponding concentration data point onto the corresponding concentration factor:

$$Y_f = B_f \, X_f \qquad\qquad [65]$$

where:

Y_f is the projection of a single concentration data point onto the fth concentration factor.

X_f is the projection of the corresponding spectral data point onto the fth spectral factor.

B_f is the proportionality constant for the fth pair of concentration and spectral factors.

Recall that the projections are often called the scores. Thus, another way of expressing equation [65] is "the scores of the spectral data points are directly proportional to the scores of the corresponding eigenvector of the concentration data points."

Figure 73 contains plots of the projections of the spectral data onto each spectral factor vs. the corresponding projections of the concentration data onto each concentration factor.

The perfectly linear, noise-free relationship between the projections is readily apparent. The slope of each relationship is equal to each proportionality constant B_f in equation [65]. B_f is sometimes called the *inner relationship*. The sign of the slope depends on the relative signs of the spectral factor vs. its corresponding concentration factor.

Next, we consider what happens when there is noise on both the absorbances and the concentration values. Figures 74 and 75 contain plots of the spectral data with noise added. Figure 76 contains plots of the concentration data with noise added. We can see that the spectral and concentration data points are no longer exactly congruent. This is because the noise in the spectral data is independent from the noise in the concentration data. Thus, in general, the noise will shift each spectral data point a different distance in a different direction than its corresponding concentration data point is shifted.

When we calculate the eigenvectors for the two different data spaces (concentration and spectral spaces) we find the corresponding spectral and concentration vectors are shifted by different amounts in different directions. This is a consequence of the independence of the noises in the concentration and spectral spaces. So, just as the noise destroyed the perfect congruence between the noise-free spectral and concentration data points, it also destroyed

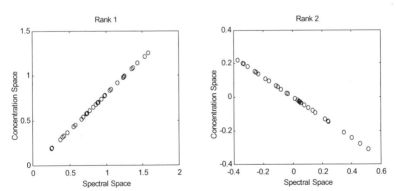

Figure 73. Projections of the concentration data onto each concentration factor vs. the corresponding projections of the spectral data onto each spectral factor for the noise-free, perfectly linear data.

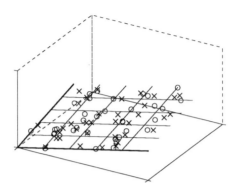

Figure 74. The spectral data from Figure 66 before the addition of noise (x) and after the addition of noise (o).

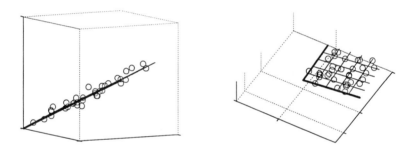

Figure 75. The spectral data from Figure 66 after the addition of noise.

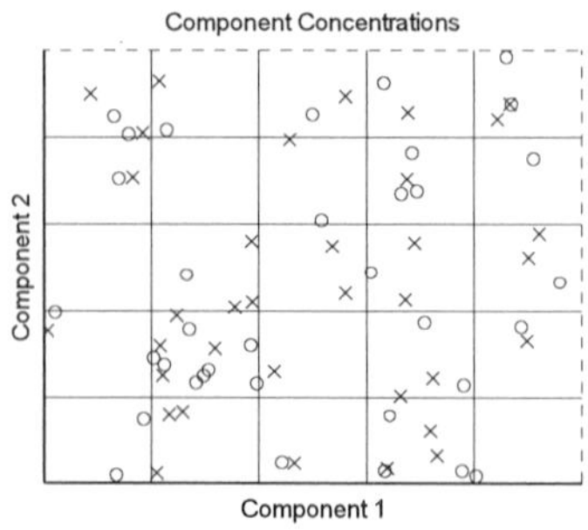

Figure 76. The concentration data from Figure 68 before the addition of noise (x) and after the addition of noise (o).

the congruence between the spectral and concentration basis vectors. It turns out that, with this particular data, the shifts of the vectors are relatively small. At the scale of our plots, the eigenvectors for the noisy data are almost identical to the noise-free eigenvectors in Figures 70 through 73. But, we will plot the projections of the spectral data onto each spectral factor vs. the corresponding projections of the concentration data onto each concentration factor. This plot may be found in Figure 77.

We see, in Figure 77, that the noise in the data impacts the relationship between the projections of the spectral data onto the each spectral factor vs. the corresponding projections of the concentration data onto each concentration factor.

Rotation of the PCA Factors by PLS

The whole idea behind PLS is to try to restore, to the extent possible, the optimum congruence between the each spectral factor and its corresponding concentration factor. For the purposes of this concept, optimum congruence is defined as a perfectly linear relationship between the projections, or scores, of the spectral and concentration data onto the spectral and concentration factors as exemplified in Figure 73. Since the spectral noise is independent from the concentration noise, a perfectly linear relationship is no longer possible. So, the best we can do is restore optimum congruence in the least-squares sense.

PLS attempts to restore optimal congruence between each spectral factor and its corresponding concentration factor by rotating them towards each other until the angle between them is zero. In other words, PLS will search for a single

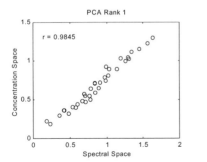
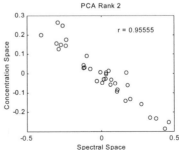

Figure 77. Projections of the concentration data onto each concentration factor vs. the corresponding projections of the spectral data onto each spectral factor for the noisy data using eigenvectors as factors.

vector, **W**, that represents the best compromise between the spectral factor and the concentration factor. The best compromise is not necessarily the factor that lies exactly half-way between the PCA factor for the spectral space and the corresponding PCA factor for the concentration space. It is, instead, the factor that maximizes (in a least-squares sense) the linear relationship between the projections (scores) of the spectral points onto the factor and the projections (scores) of the corresponding concentration points onto this same factor. In other words, PLS tries to find the factor, **W**, (for each rank, or dimensionality, of the data) that maximizes the covariance of the spectral scores with the concentrations. Each vector, **W**, will have as many elements as there are wavelengths in the spectra. By convention, **W** is usually organized as a column vector. Even though the vector, **W**, is actually an abstract factor, the elements of **W** are usually called the *loading weights*, or simply the weights.

PLS finds these factors, **W**, one-by-one. First, the most significant optimum factor, $\mathbf{W}_1$, is found. Then, that portion of the variance in the spectral data that is spanned by $\mathbf{W}_1$ is removed from the spectra. Similarly, that portion of the variance in the concentrations that is spanned by $\mathbf{W}_1$ is removed from the concentrations. Then the next factor, $\mathbf{W}_2$, is found for the spectral and concentration residuals that were not spanned by $\mathbf{W}_1$. The process is continued until all possible factors have been found.

In general, because the noise in the concentration data is independent from the spectral noise, each optimum factor, **W**, will lie at some angle to the plane that contains the spectral data. But we can find the projection of each **W**, onto the plane containing the spectral data. These projections are called the *spectral factors*, or *spectral loadings*. They are usually assigned to the variable named **P**. Each spectral factor **P**, is usually organized as a row vector.

Similarly, each optimum factor, **W**, will lie at some angle to the plane that contains the concentration data. But we can find the projection of each **W**, onto the plane containing the concentration data. These projections are called the *concentration factors*, or *concentration loadings*. They are usually assigned to the variable named **Q**. Each concentration factor **Q**, is usually organized as a row vector.

If, as in this case, all, or nearly all, of the spectral variance is linearly correlated to the concentration variance, the optimum PLS factors, **W**, and the corresponding PLS spectral factors, **P**, will tend to be very similar to each other. And **W** and **P** will, in turn, tend to be very similar to the PCA spectral factors. If, on the other hand, there is a significant amount of spectral variance that is not correlated to the variance in the concentrations, **W** and **P** will tend differ significantly from each other and from the PCA spectral factors.

When we perform PLS on the data in Figure 74, we find that the difference between the PCA factors in Figures 70 through 72 and the PLS factors was so slight that there is no point in plotting the PLS factors in separate figures. Plots of the projections onto the PLS spectral factors vs. the projections onto the PLS concentration factors are shown in Figure 78.

Let's compare these plots to the plots in Figure 77. There are, essentially, no differences between the scores (projections) on the first eigenvectors (rank 1) shown in Figure 77 and the scores on the first PLS factors shown in Figure 78. The correlation coefficient, r, is identical for the two cases. We can see some slight differences between the results for the second eigenvectors and the second PLS factors. The points in Figure 78 appear to be slightly less scattered that the points in Figure 77. Accordingly, the correlation coefficient, r, is also slightly larger for the relationship between the PLS projections.

As we've said, whether the differences between the eigenvectors and the PLS factors will be large or small is very dependent on the data itself. In this case, the relationship between the absorbance and concentration data is so strong and so linear, that there is very little that PLS can do to improve things. If, on the other hand, our data should have a large amount of variance that is unrelated to the concentrations, particularly if it is nonlinear in nature, PLS will generally succeed in efficiently rejecting it from the earlier factors. This can "free-up" degrees of freedom which can be used to enhance the regression and improve the performance of the calibration.

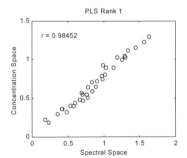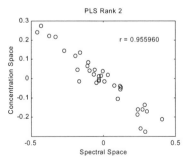

Figure 78. Projections of the concentration data onto each concentration factor vs. the corresponding projections of the spectral data onto each spectral factor for the noisy data using the PLS factors.

PLS can be counterproductive for data that has significant levels of nonlinearities that are *systematically* related to the concentrations. PLS will generally reject such nonlinearities from the earlier factors even though they can have predictive value. PCR can often produce better calibrations that PLS from data of this type. Also, PCR can usually produce better calibrations than PLS if there are large errors in the training set concentrations, particularly if the concentration errors have a systematic component. In such cases, PLS can tend to overfit the concentration errors. This type of overfitting can reduce the generality of the calibration resulting in larger errors when it is used to predict the concentrations in independent unknowns.

The point is that it is usually advisable to generate calibrations using *both* PCR and PLS. We can then evaluate each calibration validation samples and choose whichever one works best in the particular application. Fortunately, most of the software packages available today make it an easy matter to quickly generate both calibrations.

PLS in Action

Now, we are ready to apply PLS to our simulated data set. For each training set absorbance matrix, **A1** and **A2**, we will find all of the possible PLS factors. Then, we will decide how many to keep as our basis set. We will use this basis set to produce calibrations that we will use to predict the concentrations of the samples in our validation sets.

We will name the PLS spectral factors calculated for training sets 1 and 2 PLSP1 and PLSP2, respectively. Similarly we will name the PLS concentration factors, loading weights, and inner relationships PLSQ1 and PLSQ2, PLSW1 and PLSW2, and PLSB1 and PLSB2, respectively.

Choosing the Optimum Rank

Just as we did for PCR, we must determine the optimum number of PLS factors (rank) to use for this calibration. Since we have validation samples which were held in reserve, we can examine the Predicted Residual Error Sum of Squares (PRESS) for an independent validation set as a function of the number of PLS factors used for the prediction. Figure 54 contains plots of the PRESS values we get when we use the calibrations generated with training sets **A1** and **A2** to predict the concentrations in the validation set **A3**. We plot PRESS as a function of the rank (number of factors) used for the calibration. Using our system of nomenclature, the PRESS values obtained by using the calibrations from **A1** to predict **A3** are named PLSPRESS13. The PRESS values obtained by using the calibrations from **A2** to predict the concentrations in **A3**

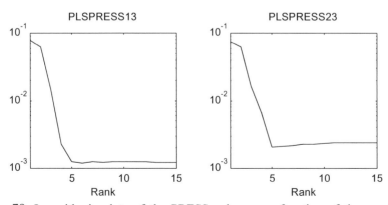

Figure 79. Logarithmic plots of the PRESS values as a function of the number of factors (rank) used to construct the calibration.

143

are named PLSPRESS23. It is clear from the plots that the errors in predicting the concentrations of the validation set, **A3**, are minimized by using 5 PLS factors for the calibration.

If we did not have a validation set available to us, we could use cross-validation for the same purposes. Figure 80 contains plots of the results of cross validation of the two training sets, **A1** and **A2**. Since no validation data is involved, we name the results PLSCROSS1 and PLSCROSS2, respectively.

Again, it is clear from these plots that the errors are minimized when 5 factors are used. Thus, we will construct our calibration matrices using a basis space cromprised of the first 5 eigenvectors (factors).

The plots in Figure 81 complete the story. They show why, if we do not have validation samples available, we cannot simply use the fits to the training set to determine how many factors to keep. First let's recall that, by fits to the training set, we mean the procedure whereby we generate a calibration from a training set and use that calibration to predict the concentrations of the samples in that same training set. We then examine the PRESS for these predictions for an indication of how well the calibration was able to fit the data in the training set. Figure 81 contains plots of the fits for the two training sets, **A1** and **A2**. Since there are no independent validation sets are involved, we have named the results PLSPRESS1 and PLSPRESS2, respectively.

When we examine the plots we see that the PRESS decreases each time we add another factor to the basis space. When all of the factors are included, the PRESS drops all the way to zero. Thus, these fits cannot provide us with any information about the dimensionality of the data. The problem is that we are

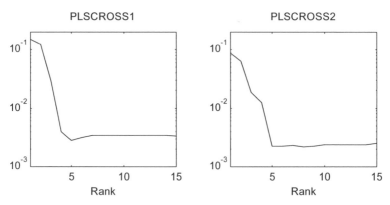

Figure 80. Logarithmic plots of the cross-validation results as a function of the number of factors (rank) used to construct the calibration.

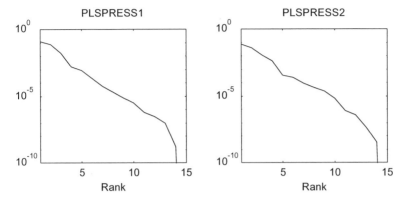

Figure 81. Logarithmic plots of the PRESS values as a function of the number of factors (rank) using the same samples for calibration and validation. As factors are added, the errors continue to decrease. When all of the factors are used, the errors equal exactly zero.

trying to use the same data for both the training and validation data. We lose the ability to assess the optimum rank for the basis space because we do not have independent validation samples that contain independent noise. So, the more PLS factors we add, the better the calibration is able to model *the particular noise* in these samples. When we use all of the factors, we are able to model the noise *completely*. Thus, when we predict the concentrations for these exact same samples that contain exactly the same noise, our PRESS values decrease each time a factor is added. When all of the factors are used, the noise is modeled completely, and the PRESS values drop to zero. So if we are trying to find the correct number of factors to use, and we do not have independent validation samples, we must use a technique such as cross-validation. Simple fits to the training set are useless for this purpose.

For our discussions, we have been using PLS to generate calibrations for all components simultaneously. Unlike PCR, it can often be advantageous to generate PLS calibrations for one component at a time. This allows PLS to find the best compromise factors for each individual component by ignoring the compromises that would be needed to accomodate the other components. When PLS is used to calibrate multiple components simultaneously, it is often called PLS-2. When used to generate calibrations for one component at a time it is often called PLS-1.

Basis Vectors and Noise Vectors

Cross-validation and PRESS both indicate that we should use 5 factors for our calibrations. These factors are the *basis factors* comprising the *basis space* for our calibration. The factors which we discard are the *noise factors*.

In the earlier chapter on PCR, we saw that we could plot each PCA factor as if it were a spectrum. The same is true for the PLS spectral factors. Figure 82 contains plots of the first 4 factors of the two training sets **A1**, and **A2**. We have named the matrices holding these factors **Vc1** and **Vc2**, respectively.

As expected, in Figure 82, we notice that the first factors for each training set are quite similar to each other and to the corresponding PCR factors we saw earlier. Again, they do not look all that abstract. Since we did not scale or center the spectra in the training sets prior to analysis, the first factor for each absorbance matrix is the least-squares average of all the spectra in the set. The second factors, on the other hand, appear a bit more abstract, although they contain spectral like features. In fact, they look a bit like first derivative spectra. This is to be expected since the second factors are capturing the maximum amount of the residual variance that correlates well with the concentrations. This residual variance is basically the least-squares average of the differences between all of the spectra and the first factor in each data set. So it should be no surprise that the second factors have these kinds of features. It should also be expected that these factors will begin to look different from each other, because the factors are now starting to pick up the individual differences between the spectra in **A1** and those in **A2**.

Moving down to the third factors, we still see spectral like features, but now the differences between the factors for each data set are becoming more noticeable. The fourth factors are more different from each other than were the first three. They still contain strong, spectral-like features, but they look a bit noisier that the earlier factors. So far, all of these factors look like they could be spanning significant information. Let's look at a few more factors which we have plotted in Figure 83.

In Figure 83 the fifth factors appear quite noisy. Nonetheless, we can imagine that there are still some systematic features in these factors. It can be dangerous to decide whether or not to retain a factor by inspecting it visually. It is too easy to see patterns in data that is really random. But, we based our

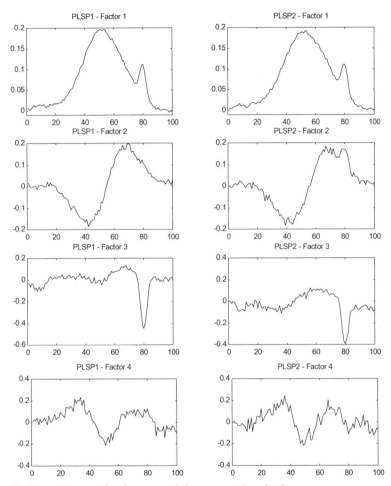

Figure 82. First 4 factors for the two training sets, **A1** and **A2**.

decision to retain this factor on the fact that we got the lowest PRESS and cross-validation values with 5 factors. The fact that we can see features on this factor serves to increase our confidence in the decision.

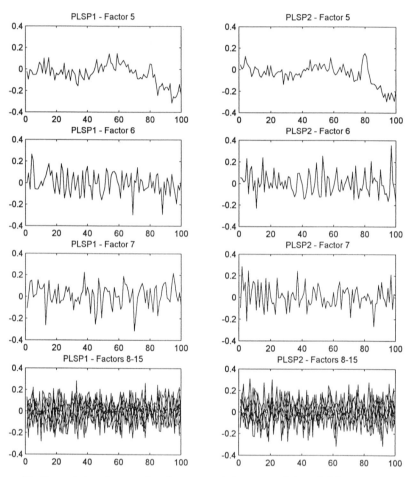

Figure 83. Remaining factors for the two training sets, **A1** and **A2**.

All of the remaining factors, 6 through 15, do appear to contain nothing but noise. Remember that true noise eigenvectors will lie in some random direction that is devoid of any useful information. Thus, barring an unusual coincidence each of them should look like pure noise when plotted in this fashion.

Regenerated Data and Residuals

Just as we did for PCR, we can use the PLS basis factors to regenerate the data. We've seen that the data regeneration step is implicit in the calibration. Even though there is no need to explicitly regenerate the data, it is, nonetheless instructive. Let's use the 5 basis vectors for training set **A1** to regenerate the spectra in **A1**. Let's also look at the residuals, that portion of the variance that is discarded from the regenerated data because it is (hopefully) pure noise. We will name the matrix holding the regenerated spectra and the residuals PLSREG1 and PLSRESID1, respectively. Figure 84 contains a plot of one of these regenerated spectra together with a plot of the original data for the same spectrum. We can easily see that a significant amount of noise has been removed without any evident degradation of the spectrum.

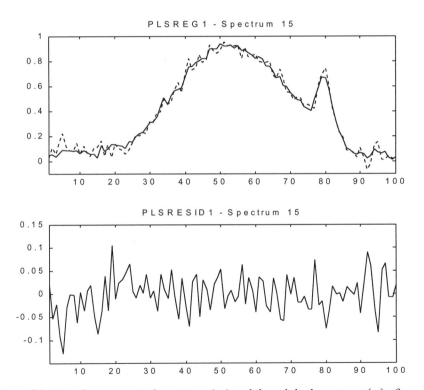

Figure 84. Plot of a regenerated spectrum (—) and the original spectrum (---) of a sample in training set A1 together with a separate plot of the differences between the two spectra.

Figure 84 also contains a plot of the differences between the original and the regenerated spectrum. This is identical to the residual spectrum. The residuals of this spectrum look comfortably like pure random noise.

Figure 85 contains plots of the residuals for all of the spectra in **A1** and **A2**. Notice that, unlike the case for PCR, these residual spectra do appear to have a small amount of spectral-like features. The features appear in the spectral regions where the most intense peaks are located. Recall that PLS chose its basis factors to optimize the *linear* regressions between the spectral factors and their corresponding concentration factors. But these spectra contain some nonlinearity. We have noted that PLS will tend to reject nonlinearities into the later factors. In this case some of the nonlinearity is spanned by noise factors which we have not included in the calibrations. Since the spectral nonlinearities in our data are strongest in the spectral regions where the spectral absorptions are strongest, it should not be surprising that it is easiest to see evidence of the rejected nonlinearities in those regions of strongest spectral absorption.

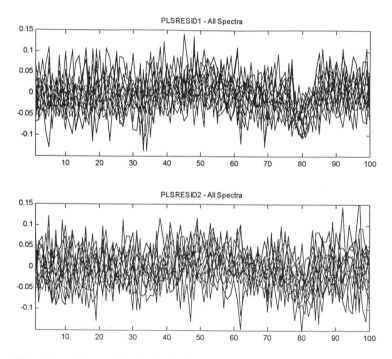

Figure 85. Plot of the residuals of all the regenerated spectra in the two training sets, **A1** and **A2**.

Whether this tendency of PLS to reject nonlinearities by pushing them onto the later factors which are usually discarded as noise factors will improve or degrade the prediction accuracy and robustness of a PLS calibration as compared to the same calibration generated by PCR depends very much upon the specifics of the data and the application. If the nonlinearities are poorly correlated to the properties which we are trying to predict, rejecting them can improve the accuracy. On the other hand, if the rejected nonlinearities contain information that has predictive value, then the PLS calibration may not perform as well as the corresponding PCR calibration that retains more of the nonlinearities and therefore is able to exploit the information they contain. In short, the only sure way to determine if PLS or PCR is better for a given calibration is to try both of them and compare the results.

We can also use the 5 factors which comprise the basis space of **A1** to regenerate the spectra in our three validation sets **A3**, **A4**, and **A5**. We will name the matrices holding these spectra, PLSREG13, PLSREG14, PLSREG15 and PLSRESID13, PLSRESID14, and PLSRESID15, respectively. Figure 86 contains plots of one regenerated spectrum from each validation set together with the same spectrum before regeneration. Figure 86 also contains plots of the residuals of all of the regenerated spectra in these validation sets.

In Figure 86, we can see that the basis space of our training set does a fine job of regenerating the validation spectra in **A3**. Noise is nicely removed without any significant degradation of the spectrum. Notice that the residuals for **A3** seem to contain some spectral features in the regions of the most intense peaks. Also notice that the residuals are roughly the same magnitude as the residuals for **A1**.

Regeneration also seems to work well on the validation spectra in **A4**. Again, we see good noise removal without spectral degradation. Notice that the regenerated spectral peak around data point 80 is somewhat different in intensity than the original spectrum. This is consistent with PLS's tendency to reject some of the spectral nonlinearities. Also, notice that the residuals for **A4** are a bit larger in magnitude than those of **A1**. This makes sense when we remember that the samples in **A4** are the overrange samples. Thus, they are somewhat different from the samples in the training set that was used to develop the basis space. So, it makes sense that the residuals are a bit higher.

As with PCR, the story is very different for the samples in **A5**. Here, we see that the regenerated spectrum has major differences from the original spectrum. Also, the residuals are much larger and display significant amounts of spectral-like features. Of course, the reasons for this are simple. **A5** contains

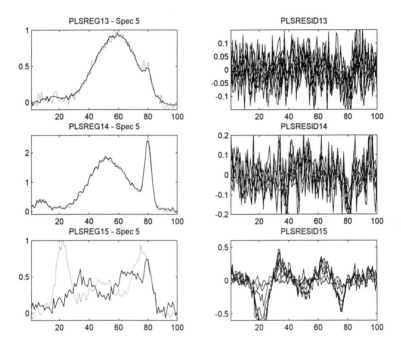

Figure 86. Plot of a regenerated spectrum (—) and the original spectrum (⋯) of a sample from each of the 3 validation sets **A3**, **A4**, and **A5**. The residuals for all of the regenerated spectra in the 3 validation sets.

samples with varying amounts of an additional, interfering component. The samples in our training set, **A1**, do not contain any of this interfering component.

So there is no way that the basis vectors for the **A1** spectra can span *all* of the variance added to the **A5** spectra by that component. Thus, it makes sense that major spectral features are missing from the regenerated spectra and show up, instead, in the residuals.

Confidence Indicator

As was the case for PCR, we see that the PLS spectral residuals for a sample will be higher whenever there is something in the data that introduces a mode of variation into the spectrum that was not present in any of the training samples used to develop the basis space. The anomolous variation could be caused by instrument drift, an unexpected interfering component, a misaligned sample cell, or whatever. We can use this property of residuals as an indicator that can signal cell, or whatever. We can use this property of residuals as an indicator that can signal us whenever a sample is significantly different from the training set samples. This is very valuable because if we try to predict the concentrations of a sample that differs significantly from the samples with which the calibration was generated, the reliability of the predictions is very poor.

We can use the sum of the squares of the residuals (SSR) of the training set as our benchmark. Then, we can establish one or more confidence limits based on this benchmark. Typically, we might set a warning level at 2 to 3 times the training set SSR. Anytime the residuals of an unknown spectrum exceed the warning level we could take appropriate action. We might turn on a yellow light, issue a warning message, send an e-mail to the person responsible for the analysis, repeat the measurement, capture a sample, save a spectrum to disk, initiate a self-diagnostic routine for the analyzer, or whatever. We could also set an alarm level. Typically this would be set at 3 to 4 times the training set SSR. If the SSR of an unknown exceeded the alarm level we could turn on a red light, sound an alarm, save the data to disk, capture a sample, initiate self-diagnostics, refuse to report the predicted concentration values, or shut down the analyzer. Table 14 shows the PLS SSR[rs] for training set **A1**, and the three validation sets, **A3** through **A5**.

Suppose we set our warning and alarm levels at typical levels of 3 and 5 times the training set SSR, respectively. We can see in Table 14 that our green light would stay on while predicting the samples in the normal validation set, **A3**. If we encounter samples from **A4**, the overrange validation set, the yellow light would come on. And when we see samples from **A5**, the validation set with the unexpected interfering component, red lights should flash, alarms should sound, etc.

Table 15 shows the SSR's for each sample in validation set **A5** together with the concentration of the unexpected component in each sample. Figure 87 contains a plot of the data in Table 15. We can see, in Table 15 that there is a

Data Set	**A1**	**A13**	**A14**	**A15**
SSR	0.1907	0.2302	0.7272	3.7450

Table 14. Sum of the square of residuals (SSR) for PLS for **A1** and **A3** through **A5**, using the 5 basis vectors for **A1**.

Data Set	A1 SSR	A5 Conc.	A15 SSR
	0.1907	0.9880	7.7626
		0.9353	6.8191
		0.8144	5.4605
		0.7733	4.7653
		0.6074	3.1067
		0.3177	1.2264
		0.1161	0.4998

Table 15. Sum of the square of residuals (SSR) for PLS for the individual samples in the validation set, A5, using the 5 basis vectors for A1 together with the concentrations of the unexpected 5th component in the A5 samples.

monotonic relationship between the SSR and the concentration of the interfering 5th component in each sample of the validation set A5. In Figure 87 we can see that the relationship is approximately linear with the square root of the SSR. The important thing is not the linearity of the relationship, but that it exists at all and increases monotonically. It gives us a very usesful way of flagging samples which our calibration may not be able to handle properly. This capability, alone, will usually give us sufficient reason to use the factor-based techniques to develop our calibrations.

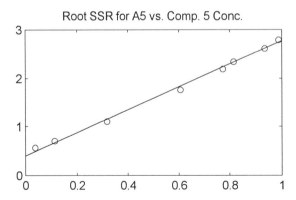

Figure 87. Semi-logarithmic plot of the PLS SSR (y-axis) vs. the concentration of Component 5 (x-axis) for each sample in A5.

PLS Calibration Matrices

Although PLS uses the regression between corresponding spectral and concentration factors to "build" a predicted concentration, factor-by-factor, it is possible to extract regression cofficients analogous to the coefficients produced by a PCR calibration by feeding the PLS calibration a "unit spectrum." We will calculate the regression coefficients for the PLS calibrations calculated from the two training sets, **A1** and **A2**. We will name these calibration matrices $PLS1_{cal}$ and $PLS2_{cal}$, respectively. In the case of PLS, that the calibration matrices have a column for each component being predicted. Each column has one regression coefficient for each spectral wavelength. Thus, we can plot each column of the PLS regression matrix as if it were a spectrum. Figure 88 contains these plots. We can think of these plots as the "strategy" of the calibration. They show us which wavelengths are taken as positively correlated with the predicted concentrations, which negatively, and which wavelengths are essentially ignored.

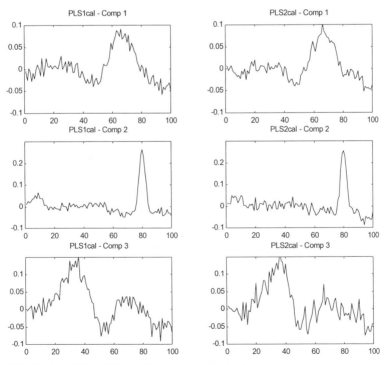

Figure 88. Plots of the PCR calibration coefficients calculated for each component with each training set.

We can see that the regression coefficients for each component produced by the two training sets are quite similar to each other. It is also apparent that the coefficients are reasonably well conditioned. In other words, their magnitude is not excessive, and they do not swing wildly from large positive to large negative values.

PLS Predictions on the Validation Sets

Finally, let's see how well the PLS calibrations predict the concentrations of our 3 validation sets **A3** - **A5**. We will use the PLS calibrations produced from each training set, **A1** and **A2**, to predict the concentrations in the validation sets. We will organize the predicted concentrations into result matrices named **PLS13**$_{res}$ through **PLS15**$_{res}$ and **PLS23**$_{res}$ through **PLS25**$_{res}$. Using this naming system, **PLS24**$_{res}$ is a concentration matrix holding the concentrations for validation set **A4** predicted with the PLS calibration matrix produced with training set **A2**. Again, there is a data "crib sheet" inside the back cover to help you keep things straight. Figure 89 contains plots of the expected vs. predicted concentrations for **PLS13**$_{res}$ through **PLS25**$_{res}$. Table 16 contains the values for PRESS, SEC^2, SEP^2, and r, for this set of results.

It is apparent that these results are essentially identical to the results obtained from this data using PCR. We even do extremely well with the overrange validation samples in **A4**. But, it would be dangerous to assume that we can routinely get away with extrapolation of this kind. Sometimes it works well, sometimes it doesn't. There is no simple rule that can tell us which situation we might be facing. It is very dependent on the particular data and

	PLS1$_{cal}$			PLS2$_{cal}$		
	PRESS	SEC^2	r	PRESS	SEC^2	r
A1	.0008	.0013	.9976	-	-	-
A2	-	-	-	.0004	.0006	.9980
A3	.0012	.0013	.9941	.0021	.0022	.9897
A4	.0035	.0032	.9987	.0061	.0068	.9962
A5	.1239	.1325	.8846	.0937	.1045	.9117

Table 16. PRESS, SEC^2, SEP^2, and r for **PLS1**$_{res}$ through **PLS25**$_{res}$.

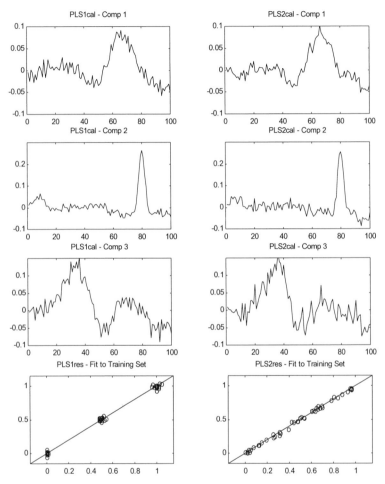

Figure 89. Expected concentrations (x-axis) vs. predicted concentrations (y-axis) for PCR calibrations (see text).

application involved. In any case, it not good practice to use a calibration to predict concentrations that fall outside the range of the concentrations that were present in the training set.

Of course, the calibrations do rather poorly predicting the concentrations of the samples in **A5**. This is exactly as expected since these samples have varying amounts of an additional, unexpected component that wasn't present in any of the calibration samples. But, with the factor-based techniques, we have the ability to detect these samples using the SSR's of the spectra. As discussed in the previous chapter on PCR, this gives us the ability to take appropriate action if we encounter any unknowns for which the calibration must be considered invalid.

The Beginning

Well, that's all there is to it! We have explored the 4 methods of quantitative analysis that are in widespread use, CLS, ILS, PCR, and PLS. We have considered the strengths and weakness of each method. The most important lesson we have (hopefully) learned is that no single method is right for all situations. Depending upon the nature of the application and the data we have to work with, any one of the 4 techniques might outperform the others. Fortunately, with the availablity of various software packages together with inexpensive computers on which to run them, it is a relatively easy matter to try them all on our data so that we may choose the method which best meets each situation.

The rest is up to you. This book includes an extensive bibliography of articles and books that can help you further master these methods. Of course, in a field as young and as active as this, bibliographies tend to be outdated the day they are compiled. New methods and new insights into existing methods appear in the literature on a regular basis. But you should now be able to read the literature critically, and with comprehension.

Very well, then, I don't see that there is any mystery about it, after all.

—Mark Twain

Appendix A: Matrices and Matrix Operations

This section will briefly review some of the basic matrix operations. It is not a comprehensive introduction to matrix and linear algebra. Here, we will consider the mechanics of working with matrices. We will not attempt to explain the theory or prove the assertions. For a more detailed treatment of the topics, please refer to the bibliography.

What is a Matrix?

For our purposes, we can simply consider a matrix as a set of scalars organized into columns and rows. For example, consider the matrix A:

$$
\mathbf{A} \;=\; \begin{bmatrix} 1 & 1 & -8 & -14 \\ 0 & 1 & -3 & -6 \\ 2 & -1 & -7 & 10 \\ 0 & 1 & -3 & -6 \\ 1 & 2 & 1 & 4 \end{bmatrix}
$$

The following statemts about **A** (or, as it is sometimes written, **[A]**) are true:

It is a 5 X 4 matrix.

The dimensions of **A** are 5 X 4.

It contains 5 rows and 4 columns.

Each row is a row vector containing 4 elements.

Each row is a 1 X 4 matrix.

Each column is a column vector containing 5 elements.

Each column is a 5 X 1 matrix.

It contains positive and negative values. (Most matrices encountered in chemometrics will contain only positive values.)

Each element of **A** is designated by using subscripts. A_{ij} denotes the element of A in the ith row and j^{th} column. For example, in the matrix **A**, above, $A_{34} = 10$.

Special Matrices

There are a number of matrices that are important enough to have special names:

Zero Matrix

As the name suggests, a zero matrix is a matrix in which all of the elements are equal to zero.

$$\begin{bmatrix} 0 & 0 & 0 & 0 \\ 0 & 0 & 0 & 0 \\ 0 & 0 & 0 & 0 \\ 0 & 0 & 0 & 0 \\ 0 & 0 & 0 & 0 \end{bmatrix}$$

Square Matrix

A square matrix is a matrix that contains the same number of columns as rows.

$$\begin{bmatrix} 0 & 1 & -3 & 6 \\ 2 & -1 & -7 & 10 \\ 0 & 1 & -3 & -6 \\ 1 & 2 & 1 & 4 \end{bmatrix}$$

Diagonal Matrix

A diagonal matrix is a square matrix in which all of the elements which do not lie on the diagonal are equal to zero. Note that the diagonal (or, more exactly, the principal diagonal) is comprised of all the elements A_{ij} for which $i = j$.

$$\begin{bmatrix} 4 & 0 & 0 & 0 \\ 0 & -1 & 0 & 0 \\ 0 & 0 & 0 & 0 \\ 0 & 0 & 0 & 4 \end{bmatrix}$$

Unit Matrix

A unit matrix is a diagonal matrix in which all of the diagonal elements are equal to 1. The unit matrix is sometimes callesd the identity matrix. It is often denoted as **I**.

$$\mathbf{I} = \begin{bmatrix} 1 & 0 & 0 & 0 \\ 0 & 1 & 0 & 0 \\ 0 & 0 & 1 & 0 \\ 0 & 0 & 0 & 1 \end{bmatrix}$$

Matrix Transpose

The transpose of a a matrix is formed by changing each column into a row (or each row into a column). The matrix transpose of a matrix, **A** is denoted by the superscript T to give $\mathbf{A}^T$.

$$\mathbf{A} = \begin{bmatrix} 1 & 1 & -8 & -14 \\ 0 & 1 & -3 & -6 \\ 2 & -1 & -7 & 10 \\ 0 & 1 & -3 & -6 \\ 1 & 2 & 1 & 4 \end{bmatrix} \qquad \mathbf{A}^T = \begin{bmatrix} 1 & 0 & 2 & 0 & 1 \\ 1 & 1 & -1 & 1 & 2 \\ -8 & -3 & -7 & -3 & 1 \\ -14 & -6 & 10 & -6 & 4 \end{bmatrix}$$

Notice that each element of **A** transpose, $\mathbf{A}^T_{ij}$ is equal to $\mathbf{A}_{ji}$. Also note that if the matrix **A** is an m by n matrix, then its transpose, $\mathbf{A}^T$ must be an m by n matrix.

We see that the transpose of the transpose returns the original matrix:

$$(\mathbf{A}^T)^T = \mathbf{A}$$

Matrix Multiplication

A matrix can be multiplied by a scalar, or by another matrix. When a matrix is multiplied by a scalar, each element of the matrix is simply multiplied by that scalar.

$$
A = \begin{bmatrix}
1 & 1 & -8 & -14 \\
0 & 1 & -3 & -6 \\
2 & -1 & -7 & 10 \\
0 & 1 & -3 & -6 \\
1 & 2 & 1 & 4
\end{bmatrix}
\qquad
2A = \begin{bmatrix}
2 & 2 & -16 & -28 \\
0 & 2 & -6 & -12 \\
4 & -2 & -14 & 20 \\
0 & 2 & -6 & -12 \\
2 & 4 & 2 & 8
\end{bmatrix}
$$

In order to multiply two matrices, the number of columns of the first matrix must be equal to the number of rows in the second matrix. If matrix A is an m X p matrix and matrix B is a p X n matrix, then they may be multiplied together to yield a matrix, C, with m rows and n columns. Each element of C is given by

$$
C_{ik} = \sum_{j=1}^{n} A_{ij} B_{jk}
$$

For example:

$$
\begin{bmatrix}
-1 & 5 \\
2 & 1 \\
\mathbf{1} & \mathbf{3}
\end{bmatrix}
\begin{bmatrix}
0 & 2 \\
1 & 4
\end{bmatrix}
=
\begin{bmatrix}
5 & 18 \\
1 & 8 \\
3 & \mathbf{14}
\end{bmatrix}
$$

We have used bold characters to show how the C_{32} was calculated. C_{32} is the dot product of the second column of matrix B with the third row of matrix A:

$$
(2 \times 1) + (4 \times 3) = 14
$$

Orthorgonality

If the dot product of two vectors is equal to zero, those vectors are orthogonal (perpendicular) to each other. For example, the dot product of the vectors:

$$\begin{bmatrix} 2 & 3 & 7 \end{bmatrix} \bullet \begin{bmatrix} -2 \\ -1 \\ 1 \end{bmatrix} =$$

$$(2 \times -2) + (3 \times -1) + (1 \times 7) = 0$$

Therefore, these two vectors must be orthogonal.

Multiplication by the Unit and Zero Matrices

From the definition of matrix multiplication, we see that the product of any matrix multiplied with a properly dimensioned zero matrix must be a zero matrix. We also see that the any matrix that is multiplied with a properly dimensioned unit matrix will remain unchanged by the multiplication.

Properties of Matrix Operations

Associative Law of Multiplication

$$A (B C) = (A B) C$$

Transpose of a Product

$$(A B)^T = B^T A^T$$

There is **NO** Commutative Law of Multiplication! Therefore, in general:

$$A B \neq B A$$

Matrix Inverse of a Square Matrix

If a square matrix has an inverse, the product of the matrix and its inverse equals the unit matrix. The inverse of a matrix A is denoted by A^{-1}.

$$A \, A^{-1} = I$$

$$A^{-1} \, A = I$$

Appendix B: Errors: Some Definitions of Terminology

It is unfortunate that the nomenclature used to describe errors in the regression steps and the prediction steps of the chemometric techniques has been a source of much confusion. Although there is general agreement on the underlying theory and practice of discussing and comparing errors, differing terminologies have been brought to bear. Even worse, some terms are used differently by different authors.

This section will briefly review some of the basic terms used to discuss errors. It is not intended to be a comprehensive treatment of the topic. Here, we will simply consider the basic definitions. We will not attempt to derive or even explain the underlying theory. For more detailed treatments, please refer to the bibliography.

What Do We Mean By Error?

For the purposes of this section, error is simply the difference between the value of the y variable predicted by a regression and the true value (sometimes called the expected value). Naturally, it is impossible to know the true value, so we are forced to settle for using the best available referee value for the y variable. (Note: it is possible that the "best available referee values" can have larger errors than the predicted values produced by the calibration.) We will follow the common convention and name the expected value of the variable y and the predicted value of the variable $\hat{y}$, pronounced "Y-hat." Then the error is given by $\hat{y} - y$. We will also denote the number of samples in a data set by the letter n.

Bias

When dealing with more than one sample, we can define the bias of a regression as the mean of the errors. This can be written as

$$\overline{\hat{y} - y} \quad .$$

PRESS

The Predicted Residual Error Sum of Squares (PRESS) is simply the sum of the squares of all the errors of all of the samples in a sample set.

$$PRESS = \sum_{i=1}^{n} (\hat{y}_i - y_i)^2$$

Many people use the term PRESS to refer to the result of leave-one-out cross-validation. This usage is especially common among the community of statisticians. For this reason, the terms PRESS and cross-validation are sometimes used interchangeably. However, there is nothing inate in the definition of PRESS that need restrict it to a particular set of predictions. As a result, many in the chemometrics community use the term PRESS more generally, applying it to predictions other than just those produced during cross-validation.

In this book, the term PRESS is used only for the case where the calibration was generated with one data set and the predictions were made on an independent data. The term CROSS is used to denote the PRESS computed during cross-validation. This was done to in an attempt to distinguish cross-validation from other means of validation.

Notice also, that PRESS, as defined, is not standardized to any absolute frame of reference. The more samples we have in our data set, the more errors there are to be squared and summed, and the larger PRESS is likely to be. Thus, PRESS is only useful for comparisons within a given data set.

Variance of Prediction

The variance of prediction, s^2, for a set of samples is defined as

$$\frac{\sum_{i=1}^{n} (\hat{y}_i - y_i - bias)^2}{n - 1}$$

As is the case for PRESS, the variance of prediction can be calculated for predictions made on independent validation sets as well as predictions made on the data set which was used to generate the calibration.

SEP

The Standard Error of Prediction (SEP) is supposed to refer uniquely to those situations when a calibration is generated with one data set and evaluated for its predictive performance with an independent data set. Unfortunately, there are times when the term SEP is wrongly applied to the errors in predicting y variables of the same data set which was used to generate the calibration. Thus, when we encounter the term SEP, it is important to examine the context in order to verify that the term is being used correctly. SEP is simply the square root of the Variance of Prediction, s^2. The RMSEP (see below) is sometimes wrongly called the SEP. Fortunately, the difference between the two is usually negligible.

MSEP

The Mean Squared Error of Prediction (MSEP) is supposed to refer uniquely to those situations when a calibration is generated with one data set and evaluated for its predictive performance with an independent data set. Unfortunately, there are times when the term MSEP is wrongly applied to the errors in predicting y variables of the same data set which was used to generate the calibration. Thus, when we encounter the term MSEP, it is important to examine the context in order to verify that the term is being used correctly. MSEP is simply PRESS divided by the number of samples.

$$MSEP = PRESS / n$$

RMSEP

The Root Mean Standard Error of Prediction (RMSEP) is simply the square root of the MSEP. The RMSEP is sometimes wrongly called the SEP. Fortunately, the difference between the two is usually negligible.

MSEE, MSEC

The Mean Squared Error of Estimate (MSEE) is sometimes called the Mean Squared Error of Calibration (MSEC). It is supposed to refer uniquely to those situations when a calibration is generated with a data set and evaluated for its predictive performance on that same data set. Unfortunately, there are times when the term MSEC is wrongly applied to the errors in predicting the y variables for a data set which is independent from the set used to generate the calibration. Thus, when we encounter these terms, it is important to examine the context in order to verify that they are being used correctly. MSEC is simply PRESS divided by the number of degrees of freedom (d.o.f.).

$$MSEP = PRESS / (d.o.f.)$$

So the trick is to understand the correct number to use for d.o.f.

For CLS, the number of degrees of freedom is equal to the number samples, n, minus the number of columns, w, in the k-matrix minus 1.

$$d.o.f. = n - w - 1$$

For ILS, the number of degrees of freedom is equal to the number of samples minus the number of wavelengths, w, used in the calibration, (i.e. the number of columns in the P matrix) minus 1.

$$d.o.f. = n - w - 1$$

For PCR the number of degrees of freedom is equal to the number of samples, n, minus the number of factors, f, used for the basis space minus 1.

$$d.o.f. = n - f - 1$$

PLS is more complex than PCR because we are simultaneously using degrees of freedom in both the x-block and the y-block data. In the absence of a rigourous derivation of the proper number of degrees of freedom to use for PLS a simple approximation is the number of samples, n, minus the number of factors (latent variables), f, minus 1.

$$d.o.f. = n - w - 1$$

It is important to emphasize that MSEC is only an indication of how well the regression was able to fit the calibrations data set. It is a major blunder to use this statistic as an indication of how well a calibration should perform when predicting samples independent from those used in the calibration set.

SEE, SEC, RMSEE, RMSE

The Standard Error of Estimate (SEE), the Standard Error of Calibration (SEC), the Root Mean Squared Error of Estimate (RMSEE), and the Root Mean Square Error (RMSE) are used interchangeably. As with MSEE, they are supposed to refer uniquely to those situations when a calibration is generated with a data set and evaluated for its predictive performance on that same data set. Unfortunately, there are times when these terms are wrongly applied to the errors in predicting y variables of a data set which is independent from the data set used to generate the calibration. Thus, when we encounter these terms, it is important to examine the context in order to verify that the term is being used correctly. SEE, SEP, RMSEE, and RMSE are simply the square root of MSEE.

Appendix C: Centering and Scaling

This section will review the basic operations for centering, weighting, and scaling data sets. We will simply review the mechanics of each operation. For a more detailed treatment of the topics, please refer to the bibliography.

There are two basic kinds of centering and scaling. Data can be treated variable by variable, or they can be treated sample by sample. For example, if we are dealing with a system of absorbance spectra measured on samples with containing two components, a variable by variable operation would deal with one component at a time, or one wavelength at a time while a sample by sample operation would deal with one spectrum or one sample at a time.

Mean Centering

Mean centering is simply an adjustment to a data set to reposition the centroid of the data to the origin of the coordinate system. From the statistical point of view, this centering is intended to prevent data points at one edge of the centroid of the data from having more influence (or leverage) than data points elsewhere on the perimeter of the centroid. Also, mean centering will remove one degree of freedom from the data and allow the calibration to focus on the differences among the points in the data set. From the analytical chemistry point of view, mean centering maps the data set into an abstract space whose origin no longer has any external physical or chemical significance. Depending upon the data and the application, mean centering can have either helpful, harmful, or neutral effect upon the performance of a calibration.

Mean centering is performed on a variable by variable basis. In other words, we would mean center a set of absorbance spectra on a wavelength by wavelength basis. Starting with the first wavelength, we compute the mean absorbance over all of the samples at that wavelength. We then subtract this mean from the absorbance value at this wavelength measured in each spectrum. When we are finished, our absorbance matrix will now contain positive and negative numbers, and the new mean absorbance value over all of the samples at each wavelength will be equal to zero. Similarly, we would mean center the concentration data for these samples on a component by component basis. The decision whether or not to mean centering the x-block data is independent from the decision about centering the y-block data. We can decide to center either, both, or neither.

Figure C1 shows a hypothetical set of data before mean centering. Figure C2 shows the same data set after mean centering. We can imagine that this is a plot of the y data (let's call them concentration values) for a two component system. For each of the 15 samples in the data set, we plot the concentration of the first component along the x-axis and the concentration of the second

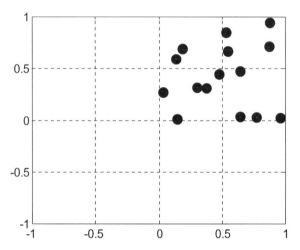

Figure C1. Hypothetical data set before mean-centering.

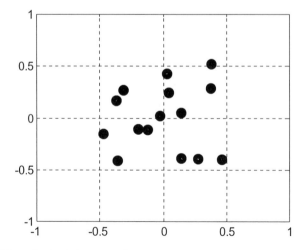

Figure C2. Hypothetical data set after mean-centering.

component along the y-axis. Note that the positions of the data points relative to one another are unchanged. The effect of mean centering has been to move the origin of the new coordinate system to the centroid of the data points.

Variance Scaling

Variance scaling is an adjustment to a data set that equalizes the variance of each variable. From the statistical point of view, this is intended to equalize the influence of each variable in the data set. Variance scaling removes one degree of freedom from the data. From the analytical chemistry point of view, variance scaling maps the data set into an abstract space whose axes no longer have any external physical or chemical significance. It also can reduce the influence of variables where the signal variation (and hence analytically useful information content) is large while increasing the influence of variables that contain mostly noise. It is becoming understood that variance scaling will usually introduce into the calibration an undesireable sensitivity to changing conditions. Generally speaking, it is usually best not to variance scale unless you have a specific reason for doing it.

Variance scaling is performed on a variable by variable basis. In other words, we would variance scale a set the concentration values of a data set on a component by component basis. Starting with the first component, we compute the total variance of the concentrations of that component. There are several variations on variance scaling. First, we will consider the most the method which adjusts all the variables to exactly unit variance. To do this we compute the variance of the variable, and then use the variance to scale all the concentrations of all the samples so that the new variance for the component is equal to unity.

To compute the variance, we first find the mean concentration for that component over all of the samples. We then subtract this mean value from the concentration value of this component for each sample and square this difference. We then sum all of these squares and divide by the degrees of freedom (number of samples minus 1). The square root of the variance is the standard deviation. We adjust the variance to unity by dividing the concentration value of this component for each sample by the standard deviation. Finally, if we do not wish mean-centered data, we add back the mean concentrations that were initially subtracted. Equations [C1] and [C2] show this procedure algebraically for component, k, held in a column-wise data matrix.

First compute the standard deviation for the variable:

$$\text{standard deviation} \quad = \quad s_k \quad = \frac{\displaystyle\sum_{i=1}^{n}(a_{ki}-\bar{a}_k)}{(n-1)} \qquad [C1]$$

Then scale each point by the standard deviation

$$a'_{ki} \quad = \quad \frac{a_{ki}-\bar{a}_k}{s_k} + \bar{a}_k \qquad [C2]$$

The decision whether or not to variance scale the x-block data is independent from the decision about scaling the y-block data. We can decide to scale either, both, or neither.

Figure C3 shows the same data from figure C1 after variance scaling. Figure C4 shows the mean centered data from figure C2 after variance scaling. Variance scaling does change the positions of the data points from one another, but does not change the location of the centroid of the data set.

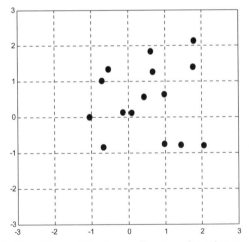

Figure C3. The data from Figure C1 after scaling to unit variance.

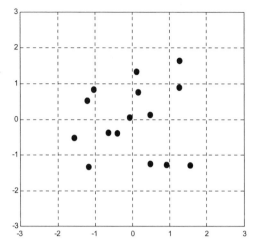

Figure C4. The data from Figure C2 after scaling to unit variance.

An alternative method of variance scaling is to scale each variable to a uniform variance that is not equal to unity. Instead we scale each data point by the root mean squared variance of all the variables in the data set. This is, perhaps, the most commonly employed type of variance scaling because it is a bit simpler and faster to compute. A data set scaled in this way will have a total variance equal to the number of variables in the data set divided by the number of data points minus one. To use this method of variance scaling, we compute a scale factor, s_f, over all of the variables in the data matrix, a_{ij},

$$s_f = \sqrt{\sum_{i=1, j=1}^{n,n} a_{ij}^2} \qquad [C3]$$

We then remove then mean center each data point, a_{ij}, and divide it by the scale factor. If we do not wish to mean-center the data, we finish by adding the mean value back to the scaled data point.

$$aij = \frac{a_{ij} - \bar{a}_i}{s_f} + \bar{a}_i \qquad [C4]$$

Figure C5 shows the data from Figure C1 after this type of scaling to uniform variance. Figure C6 shows the mean-centered data from Figure C2 after the same treatment.

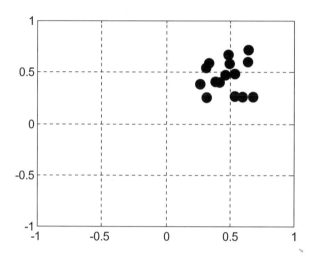

Figure C5. The data from Figure C1 after scaling to uniform variance.

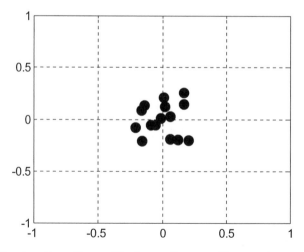

Figure C6. The data from Figure C1 after scaling to uniform variance.

Autoscaling

Autoscaling is another term that has been used in different ways by differnt people. It is often used to indicate "mean centering followed by variance scaling." Others use it to indicate normalization (see below).

Normalization

Normalization, is an adjustment to a data set that equalizes the magnitude of each sample. In other words, normalization removes all information about the distance each data point lies from the origin of the data space but preserves the direction. Normalization has a relatively limited number of special applications. For example, it is frequently used a pre-processing step in preparing reference spectra for a qualitative identification library. The idea is to retain only the information that qualitatively distinguishes one sample from another while removing all information that could separate two samples of identical composition but different concentrations.

Normalization is performed on a sample by sample basis. For example, to normalize a spectrum in a data set, we first sum the squares of all of the absorbance values for all of the wavelengths in that spectrum. Then, we divide the absorbance value at each wavelength in the spectrum by the square root of this sum of squares. Figure C7 shows the same data from Figure C1 after variance scaling. Figure C8 shows the mean centered data from Figure C2 after variance

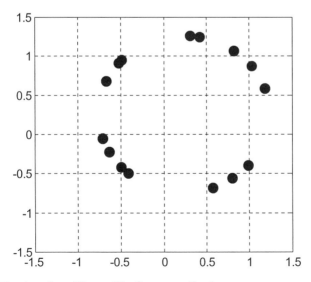

Figure C7. The data from Figure C1 after normalization.

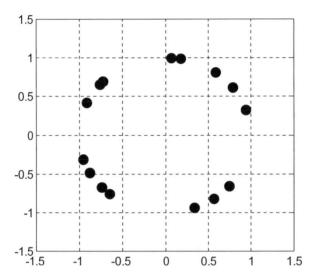

Figure C8. The data from Figure C2 after normalization.

scaling. Notice that, in both figures the data points all lie on a circle. In Figure C7 we see that, for the nonmean centered case, the circle is centered on the centroid of the original data points. In Figure C8, the mean centered data, we see that the circle is centered on the origin of the new coordinate system. The magnitude information has been eliminated. All that remains is the direction from origin. Normalization together with mean centering is sometimes called autoscaling. Note that the term autoscaling is one of those terms that is used in different ways by different people. So whenever we see the term used, it is important to investigate the context in order to understand what it means.

Appendix D: F-Test for Reduced Eigenvalues

Eigenvalues and Reduced Eigenvalues

A data set, existing in its native coordinate system, can be described in terms of an alternate coordinate system defined by the eigenvectors of the data. The first n of these eigenvectors will span the meaningful information in the data set while all of the remaining eigenvectors will span only noise. Each eigenvector has, associated with it, an eigenvalue which represents the magnitude of the total variance spanned by that eigenvector. Malinowski has shown that, if the errors in the data are uniformly distributed, each eigenvalue can be normalized for the degrees of freedom in the data with respect to that eigenvalue. The jth normalized eigenvalue Rev_j, also known as a *reduced eigenvalue*, for a data matrix with r rows and c columns is calculated by normalizing the jth eigenvalue, Ev_j

$$Rev_j = Ev_j / (r - j + 1)(c - j + 1) \qquad [D1]$$

Note that there can be a maximum of s meaningful eigenvectors, eigenvalues, and reduced eigenvalues where s is the lessor of r and c. Because the reduced eigenvalues for noise eigenvectors should all belong to the same normal distribution, we can apply an F-test to determine if a given reduced eigenvalue corresponds to a noise eigenvector which may be discarded, or to a basis vector which should be retained.

F-Test for reduced eigenvalues

We use a two-way F-test to determine if the reduced eigenvalue for the jth eigenvector is, to a chosen degree of probability, greater in magnitude than all of the reduced eigenvectors which come after it. This requires us to calculate the F statistic, F_j, for each reduced eigenvalue except the last. We can do this directly from each eigenvalue Ev_j

$$F = \frac{\sum\limits_{n=j+1}^{s} (r-n+1)(c-n+1)}{(r-j+1)(c-j+1)} \frac{Ev_j}{\sum\limits_{n=j+1}^{s} Ev_n} \qquad [D2]$$

Starting with the next to the last eigenvalue, we can then compare the F statistic calculated for that eigenvalue to the F(1, s - j) value in the statistical table of F percentage points at the desired level of significance (usually 5% or 10%). The notation $F(1, s - j)$ indicates that when consulting the tables, we always use a numerator of 1, while the denominator is $(s - j)$. If the calculated F statistic for the next to the last eigenvalue is below the value in the statistical tables, we conclude that it is, indeed, a noise eigenvalue and move to the next higher (in magnitude) eigenvalue. Note that each time we move to the next higher eigenvalue the denominator $(s - j)$ will increase by 1. At the first instance where an eigenvalue's calculated F statistic exceeds the corresponding F value in the statistical tables, we conclude that this eigenvalue, together with all of the eigenvalues higher than it, corresponds to basis eigenvectors which should be retained, and the test is complete.

Appendix E: Leverage and Influence

Leverage

Simply speaking, the leverage of a single data point is directly proportional to its distance from the origin of the data space. In general, the greater the leverage of a data point, the greater its influence on the principle components or PLS factors, as well as the slope (and in some cases the intercept) of a regression line computed with a least squares method.

Figure E1 shows a data set which contains one data point, **A**, with significantly higher leverage and one data point, **B**, with significantly lower leverage than the other points in the data set. Also shown is the first principle component for the data set.

Figure E2 shows a least-squares regression between the scores of this data set on the first principle component and a dependent variable. If the regression is constrained to pass through the origin at 0,0, it is evident that an incremental change in the slope of the regression line will cause a greater incremental change in error of the fit to point **A** than in the error of the fit to any other point in the data set. Since the regression line is chosen to minimize the sum of the squares of all of the errors, point **A** will therefore have more influence on the slope of the regression line than any other single point in the data set. If a regression method is used which allows for a regression line with a nonzero intercept, the influence of point **B** on the regression will also be larger than the

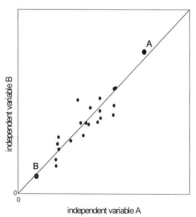

Figure E1. Hypothetical data set containing points with, **A**, atypically high and, **B**, atypically low leverage.

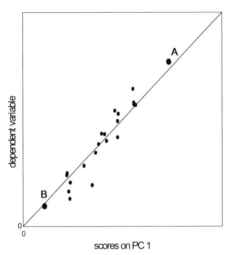

scores on PC 1

Figure E2. Hypothetical regression between a dependent variable and the scores of the data set containing points with, **A**, atypically high and, **B**, atypically low leverage.

remainder of the points. Whether the larger influence of points **A** and **B** is good, bad, or indifferent depends upon whether **A** and **B** represent good, valid data for their respective regions of the calibration space, or whether they are atypical outliers. A point can be an outlier due to unusually large measurement errors, or because it is a point which is outside the range of the normal calibration space for which the regression is required. For example, many measurement techniques involve a noise level which is independent of the magnitude of the measured value. Thus, points such as point **B** which are close to the origin of the native data space will have a poorer signal-to-noise ratio than points further from the origin.

If we mean-center the data in Figure E1 before performing the regression, we would have the situation depicted in Figure E3. Here it is evident that both points A and B will have a larger influence on the slope of the regression line than the other points. When we consider points such as point **B** often have the poorest signal-to-noise ratios of all the points in the data set, we can see that mean-centering data prior to regression can be detrimental because it can give the data points with the poorest signal-to-noise ratios more influence over the regression than points with better signal-to-noise ratios.

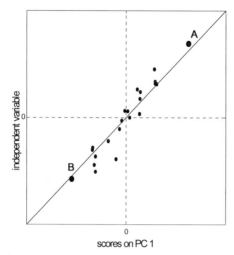

Figure E3. Hypothetical mean-centered data set containing points with, **A**, atypically high and, **B**, atypically low leverage.

Mahalanobis Distances

So far we have been considering leverage with respect to a point's Euclidean distance from an origin. But this is not the only measure of distance, nor is it necessarily the optimum measure of distance in this context. Consider the data set shown in Figure E4. Points **C** and **D** are located at approximately equal Euclidean distances from the centroid of the data set. However, while point **C** is clearly a typical member of the data set, point **D** may well be an outlier. It would be useful to have a measure of distance which relates more closely to the similarity/difference of a data point to/from a set of data points than simple Euclidean distance.the various Mahalanobis distances are one such family of such measures of distance. Thus, while the Euclidean distances of points **C** and **D** from the centroid of the data set are equal, the various Mahalanobis distances from the centroid of the data set are larger for point **D** than for point **C**.

Influence Plots

It is often helpful to examine the regression errors for each data point in a calibration or validation set with respect to the leverage of each data point or its distance from the origin or from the centroid of the data set. In this context, errors can be considered as the difference between expected and predicted (concentration, or y-block) values for the regression, or, for PCA, PCR, or PLS, errors can instead be considered in terms of the magnitude of the spectral

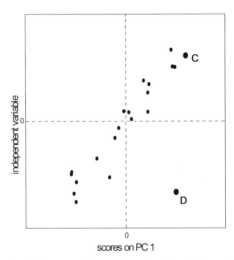

Figure E4. Hypothetical data set illustrating that Euclidean distance is not an ideal metric for membership in a data set.

(x-block) residuals for each sample. A plot of errors vs. leverage or Mahalanobis distance is often called an influence plot. Figure E5 shows an influence plot for some hypothetical data. A point with relatively low leverage but high error such as point **E** is very often an outlier due to some error in the

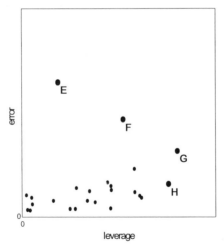

Figure E5. Influence plot for a hypothetical data set showing, **E**, a probable outlier; **F** and **G**, questionable outliers; and, **H**, a probable atypical but important calibration sample.

measurement, or to an atypical sample for which the calibration is not valid. Point **F** and **G** which have relatively high leverage and errors might also be outliers, or they might be atypical points which are important to the calibration but are much less common than the majority of the points in the data set. Point **H** which has higher leverage than the majority of data points but a typical error, is likely to be a valuable atypical point which is important to the calibration. The normal courses of action here would be to further review sample **E** to confirm the validity of excluding it from the calibration set; to understand why sample **H** is different from the other calibration points, and the implications for the calibration; and to further investigate samples **F** and **G** to determine why they are different and whether they should be discarded from the data set, retained, or remeasured.

Bibliography

General

1. "Standard Practices for Infrared, Multivariate, Quantitative Analysis", ASTM E 1655-97.

2. Beebe, K.R., Kowalski, B.R., "An Introduction to Multivariate Calibration and Analysis", *Anal. Chem.* 1987 (59) 1007A-1017A.

3. Draper, N., Smith, H., *Applied Regression Analysis*, 2nd edition, John Wiley and Sons, New York, 1981.

4. Malinowski, E.R., et. al. *Factor Analysis in Chemistry*, 2nd edition, John Wiley and Sons, New York, 1991.

5. Mark, H., *Principles and Practice of Spectroscopic Calibration*, John Wiley and Sons, New York, 1991.

6. Martens, H., Naes, T., *Multivariate Calibration*, John Wiley and Sons, New York, 1989.

7. McClure, G.L., Ed. *Computerized Quantitative Infrared Analysis*, ASTM STP 934, American Society for Testing and Materials, Philadelphia, 1987.

Multiple Linear Regression (MLR), Classical Least-Squares (CLS, K-matrix), Inverse Least-Squares (ILS, P-matrix)

8. Brown, C.W., "Classical and Inverse Least-Squares Methods in Quantitative Spectral Analysis", *Spectrosc.* 1986 (1) 23-37.

9. Brown, C.W., Lynch, P.F., Obremski, R.J., Lavery, D.S., "Matrix Representations and Criteria for Selecting Analytical Wavelengths for Multicomponent Spectroscopic Analysis", *Anal. Chem.* 1982 (54) 1472-1479.

10. Carey, W.P., Beebe, K.R., Kowalski, B.R., "Multicomponent Analysis using an Array of Piezoelectric Crystal Sensors", *Anal. Chem.* 1987 (59 1529-1534.

11. Haaland, D.M. "Classical versus Inverse Least-Squares Methods in Quantitative Spectral Analyses", *Spectrosc.* 1987 (2) 56-57.

12. Haaland, D.M., et.al. "Application of New Least-squares Methods for the Quantitative Infrared Analysis of Multicomponent Samples", *Appl. Spec.* 1982 (36) 665-673.

13. Haaland, D.M. et.al. "Improved Sensitivity of Infrared Spectroscopy by the Application of Least Squares Methods", *Appl. Spec.* 1980 (34) 539-548.

14. Haaland, D.M., et. al. "Multivariate Least-Squares Methods Applied to the Quantitative Spectral Analysis of Multicomponent Samples", *Appl. Spec.* 1985 (39) 73-84.

15. Kargacin, M.E., et. al. "Ion Intensity and Image Resolution in Secondary Ion Mass Spectrometry", *Anal. Chem.* 1986 (58) 2300-2306.

16. Kisner, H.J., et. al. "Multiple Analytical Frequencies and Standards for the Least-Squares Spectrometruc Analysis of Serun Lipids", *Anal. Chem.* 1983 (55) 1703-1707

17. H.J. Kisner, C.W. Brown, G.J. Kavarnos, "Simultaneous Determination of Triglycerides, Phospholipids, and Cholesteryl Esters by Infrared Spectrometry", *Anal. Chem.* 1982 (54) 1479-1485.

18. Lam, R.B. "On the Relationship of Least Squares to Cross-correlation Quantitative Spectral Analysis", *Appl. Spec.* 1983 (37) 567-569.

19. Maris, M.A., C.W. Brown, G.J. Kavarnos, "Nonlinear Multicomponent Analysis by Infrared Spectrophotometry", *Anal. Chem.* 1983 (55) 1694-1703.

20. McClure, G.L., et. al. "Application of Computerized Quantitative Infrared Spectroscopy to the Determination of the Principal Lipids Found in Blood Serum", *Computerized Quantitative Infrared Analysis*, ASTM STP 934, G.L. McClure, Ed. American Society for Testing and Materials, Philadelphia, 1987, 131-154.

21. Otto, M. et. al. "Spectrophotometric Multicomponent Analysis Applied to Trace Metal Determinations", *Anal. Chem.* 1985 (57) 63-69.

Principal Component Regression (PCR, PCA, Factor Analysis)

22. Antoon, M.K., et. al. "Factor Analysis Applied to Fourier Transform Infrared Spectra", *Appl. Spec.* 1979, (33) 351-357.

23. Are, H.A., Marum, P., "On the Effect of Calibration and the Accuracy of NIR Spectroscopy with High Levels of Noise in the Reference Values", *Appl. Spec.* 1991 (45) 109-115.

24. Bulmer, J.T., et. al. "Factor Analysis as a Complement to Band Resolution Techniques. I. The Method and its Application to Self-Association of Acetic Acid", *J. Phys. Chem.* 1973, (77) 256-262.

25. Culler, S.R., et. al. "Factor Analysis Applied to a Silane Coupling Agent on E-Glass Fiber System", *Appl. Spec.* 1984 (38) 495-500.

26. Dale, J.M., Klatt, L.N., "Principal Component Analysis of diffuse Near-Infrared Reflectance Data From Paper Currency", *Appl. Spec.* 1989 (43) 1399-1405.

27. Gillette, P.C., et. al. "Noise Reduction via Factor Analysis in FT-Ir Spectra", *Appl. Spec.* 1982, (36) 535-539.

28. Kargacin, M.E., et. al. "Ion Intensity and Image Resolution in Secondary Ion Mass Spectrometry", *Anal. Chem.* 1986, (58) 2300-2306.

29. Lindberg, W., et. al. "Partial Least Squares Method for Spectrofluorimetric Analysis of Mixtures of Humic Acid and Ligninsulfonate", *Anal. Chem.* 1983, (55) 643-648.

30. Lukco, R.G., Kosman, J.J., "The Use of GC-AES Multielement Simulated Distillation for Petroleum Product Fingerprinting", *J. Chrom. Sci.* 1993, March

31. Malinowski, E.R., et. al. *Factor Analysis in Chemistry*, 2nd edition, John Wiley and Sons, New York, 1991.

32. Malinowski, E.R., "Theory of the Distrubution of Error Eigenvalues Resulting from Principal Component Analysis with Applications to Spectroscopic Data", *J. Chemo.* 1987 (1) 33-40.

33. Malinowski, E.R., "Statistical F-Tests for Abstract Factor Analysis and Target Testing", *J. Chemo.* 1987 (1) 49-60

34. Malinowski, E.R. "Theory of Error in Factor Analysis", *Anal. Chem.* 1977, (49) 606-612.

35. Malinowski, E.R. "Determination of the Number of Factors and the Experimental Error in a Data Matrix", *Anal. Chem.* 1977, (49) 612-617.

36. Naes, T., Isaksson, T., "Selection of Samples for Calibration in Near-Infrared Spectroscopy. Part I: General Principles Illustrated by Example", *Appl. Spec.* 1989 (43) 328-335.

37. Rao, G.R., et. al. "Factor Analysis and Least-Squares Curve- Fitting of Infrared Spectra: An Application to the Study of Phase Transitions in Organic Molecules", *Appl. Spec.* 1984, (38) 795-803.

38. Schostack, K.J., Malinowski, E.R., "Preferred Set Selection by Iterative Key Set Factor Analysis", *Chemo. and Intel. Lab. Sys.* 1989 (6) 21-29.

39. Vaughan, M., Templeton, D.M., "Determination of Ni by ICP-MS: Correction of Oxide and Hydroxide Interferences Using Principal Components Analysis", *Appl. Spec.* 1990 (44) 1685-1689.

Partial Least-Squares in Latent Variables (PLS)

40. Carey, W.P., Beebe, K.R., Kowalski, B.R., "Multicomponent Analysis using an Array of Piezoelectric Crystal Sensors", *Anal. Chem.* 1987 (59) 1529-1534.

41. Donahue, S.M., Brown, C.W., Scott, M.J., "Analysis of Deoxyribonucleotides with Principal Component and Partial Least-Squares Regression of UV Spectra after Fourier Processing", *Appl. Spec.* 1990 (44) 407-413.

42. Geladi, P., Kowalski, B.R., "Partial Least-Squares Regression: A Tutorial", *Anal. Chim. Acta*, 1986 (185) 1-17.

43. Haaland, D.M., Thomas, E.V., "Partial Least-Squares Methods for Spectral Analysis 1. Relation to Other Quantitative Calibration Methods and the Extraction of Qualitative Information" *Anal. Chem.* 1988 (60) 1193-1202.

44. Haaland, D.M., Thomas, E.V., "Partial Least-Squares Methods for Spectral Analysis 2. Application to Simulated and Glass Spectral Data" *Anal. Chem.* 1988 (60) 1202-1208.

45. Hanna, D.A., Hangac, B.A., et. al., "A Comparison of Methods Used for the Reconstruction of GC/FT-IR Chromatograms", *J. Chrom. Sci.* 1979 (17) 423-427.

46. Kelly, J.J., Barrow, C.H., et. al., "Prediction of Gasoline Octane Numbers from Near-Infrared Spectral Features in the range 660-1215nm", *Anal. Chem.* 1989 (61) 313-320.

47. Lindberg, W., Persson, J.-A., "Partial Least-Squares Method for Spectrofluorimetric Analysis of Mixtures of Humic Acid and Ligninsulfonate" *Anal. Chem.* 1983 (55) 643-648

48. Otto, M, Wegscheider, W, "Spectrophotometric Multicomponent Analysis Applied to Trace Metal Determinations", *Anal. Chem.* 1985 (57) 63-69.

Miscellaneous and Advanced Topics

49. Frans, S.D., Harris, J.M., "Selection of Analytical Wavelengths for Multicomponent Spectrophotometric Determinations", *Anal. Chem.* 1985 (57) 2680-2684.

50. Hirschfeld, T., "Limits of Analysis", Anal. Chem. 1976 (48) 17A-31A.

51. Mahalanobis, P.C., "On the Generalized Distance in Statistics", *Proc. Natl. Inst. Sci. India*, 1936, (2) 49-55.

52. Mark, H. "A Monte Carlo Study of the Effect of Noise on Wavelength Selection during Computerized Wavelength Searches", *Appl. Spec.* 1988 (8) 1427-1440.

53. Mark, H., "Normalized distances for Qualitative Near-Infrared Reflectance Analysis", *Anal. Chem.* 1986 (58) 379-384.

54. Mark, H. "Use of Mahalanobis Distances to Evaluate Sample Preparation Methods for Near-Infrared Reflectance Analysis", *Anal. Chem.* 1987 (59) 790-795.

55. Mark, H., Norris, K., Williams, P., "Methods of Determining the True Accuracy of Analytical Methods", *Anal. Chem.* 1989 (61) 398-403.

56. Mark, H., Workman, J., *Statistics in Spectroscopy*, Academic Press, Boston, 1991.

57. Schostack, K.J., and Malinowski, E.R., "Preferred Set Selection by Iterative Key Set Factor Analysis," *Chemo. and Intel. Lab. Sys.* 1989 (6) 212-229.

Index